Dynamical Evolution of Globular Clusters

PRINCETON SERIES IN ASTROPHYSICS

Edited by Jeremiah P. Ostriker

Theory of Rotating Stars, *by J. L. Tassoul*
Theory of Stellar Pulsation, *by J. P. Cox*
Galactic Dynamics, *by James Binney and Scott Tremaine*
Dynamical Evolution of Globular Clusters, *by Lyman Spitzer, Jr.*

Lyman Spitzer, Jr.

DYNAMICAL EVOLUTION OF GLOBULAR CLUSTERS

PRINCETON UNIVERSITY PRESS
Princeton, New Jersey

Published by Princeton University Press, 41 William Street, Princeton, New Jersey 08540
In the United Kingdom: Princeton University Press, Guildford, Surrey

Library of Congress Cataloging-in-Publication Data

Spitzer, Lyman, 1914–
Dynamical evolution of globular clusters.

(Princeton series in astrophysics)
Bibliography: p. Includes index.
1. Stars—Globular clusters. I. Title. II. Series.
QB853.S66 1987 523.8′55 87-45536
ISBN 0-691-08309-6 (alk. paper)
ISBN 0-691-08460-2 (pbk.)

Most of the figures in this book have been reproduced, usually with some adaptations, from other sources. They are used here with the kind permission of the authors cited in the figure captions. Figures 1.2(b), 1.3, 4.1 and 7.4 are reproduced by courtesy of the International Astronomical Union.

This book has been composed in Linotron Times Roman

Clothbound editions of Princeton University Press books are printed on acid-free paper, and binding materials are chosen for strength and durability. Paperbacks, although satisfactory for personal collections, are not usually suitable for library rebinding

Printed in the United States of America by Princeton University Press
Princeton, New Jersey

To my colleagues

whose support and stimulation

have been invaluable

Contents

Preface

The dynamical evolution of an isolated spherical system composed of very many mass points has an appealing simplicity. The Newtonian laws of motion are exact, and all average quantities are functions only of radial distance r and time t. Nevertheless, it is only recently, with the availability of fast computers, that a systematic understanding of how such systems develop through time has emerged. Since these idealized systems should provide a very good approximation for globular clusters in this and other galaxies, the theory of their development is an important part of astronomy as well as an interesting branch of theoretical particle dynamics.

This book analyzes the various processes that occur as a globular cluster evolves dynamically. Written for the use of astronomers and physicists interested in this active research field, the book presents a deductive approach, with most of the essential results derived from first principles. The treatment is somewhat compressed, with some of the derivations sketched in outline rather than spelled out in detail, an approach which may have some educational value in developing a clearer understanding of the principles involved.

Chief emphasis is placed on those aspects which are believed to play a major role in the evolution of most clusters and which have been studied in sufficient detail to permit a coherent and somewhat realistic theory. Thus the book is focussed on the changes of stellar velocities resulting from random encounters between pairs of stars (chapter 2) and on the various evolutionary changes which result (chapters 3 and 4). A major complicating factor considered is the galactic gravitational field, which perturbs stellar orbits in the clusters moving about in this field and modifies their dynamical evolution (chapter 5). An important phase of most model clusters is the collapse of the central core; when the distances between stars become relatively short, binary stars are formed, and the role of these objects in terminating core collapse and affecting post-collapse evolution is analyzed (chapters 6 and 7).

In view of the emphasis on pure theory, no detailed comparison is given here between theoretical models and the observations. However, the introductory chapter 1 summarizes the observed properties of globular clusters, providing observational guidance for the subsequent theoretical developments, also summarized in this chapter.

The origin and early evolution of globular clusters is ignored because so little is known about this important early phase. Evolution after core collapse is discussed only briefly since our understanding of this phase is still very incomplete. Several topics are ignored because their relevance to real clusters is uncertain; these include the effect of a massive black hole at the cluster center, the dynamical instability of a cluster in which the distribution of velocities is anisotropic, and the effect of special and general relativity on clusters with random velocities comparable with the light velocity. Elegant and extensive research has been done on such topics, which are possibly relevant to galactic nuclei, but the observations do not indicate any direct connection of these subjects with globular clusters.

The references are restricted to publications which give further technical details on some of the specific subjects discussed in the book. Thus the bibliography is not in any sense complete and omits a number of papers of substantial importance in the history of this field. More complete lists of relevant papers may be found in the Proceedings of two IAU Symposia, Nos. 69 and 113, each dealing specifically with the dynamical evolution of clusters. Reference numbers here are given in square brackets, with references listed at the end of each chapter. The subject index and the directory of symbols may be helpful in the use of this volume.

This book was started in late 1984, and the early chapters include very little that was published after that time. For chapters 6 and 7 the effective cut-off date was the late spring of 1986. Ignoring important preprints that reached me during the summer of 1986 was a painful necessity.

It is a great pleasure to acknowledge the information and helpful suggestions I have received from many astronomical colleagues, including J. Binney, J. Goodman, M. Hénon, P. Hut, S. Inagaki, I. King, D. Lynden-Bell, S. McMillan and S. Shapiro. H. Cohn kindly provided me with much of the unpublished analysis summarized in §2.3. Very detailed comments on the manuscript were made by D. Heggie, A. Lightman, J. Ostriker, T. Statler and S. Tremaine. I am especially indebted to Venita Nixon, my secretary during 30 years, who has converted my jumbled drafts of several books into elegant manuscripts. Much of this book was written during a four-month stay, from December 1984 to March 1985, at the Institut d'Astrophysique, Paris, France. As in the past, I greatly appreciate the warm hospitality of the Institut and of its Director, Professor Jean Audouze.

Princeton University Observatory
August 1986

Dynamical Evolution of Globular Clusters

1

Overview

The structure of a globular cluster—the spatial distribution and random velocities of its stars—changes slowly with time as a result of simple dynamical principles. In much of this dynamical evolution, the changes in the cluster are a consequence of Newton's laws of motion and of gravitation as applied to an assembly of mass points. The theory of such changes is the general topic of this book.

The following sections in this first chapter present the observational and theoretical background for the later chapters. Relevant observations on globular clusters, especially data on the structure, age and mass of these systems, are summarized in §1.1. Little is said about most of the physical properties of the individual stars, including their composition and internal structures; these properties affect dynamical evolution of clusters only under unusual circumstances, not yet studied in detail. The subsequent sections give the theoretical point of view adopted in the rest of the book. Thus §1.2 describes the basic zero-order approximation in which the cluster is spherically symmetric, the gravitational potential, ϕ, varies smoothly with distance, r, from the cluster center, and many types of steady solutions, with no evolution, are possible. As detailed in the subsequent section §1.3, such zero-order solutions are subject to a variety of perturbations, mostly resulting from time dependent gravitational fields, including small-scale fields within the cluster and larger-scale fields produced by external masses, especially the Galaxy. The specific effects which these various perturbations can gradually produce on the structure of the cluster are also summarized in this section.

1.1 OBSERVATIONS

In common with other galaxies, our own system includes [1] some 150–200 globular clusters, each containing typically some 10^5 stars. More than half of the observed clusters are within 10 kpc of the galactic center, but the distribution extends to much greater distances, with several beyond 50 kpc. Their z distances (measured perpendicular to the galactic plane) do not differ much on the average from those in x and y.

A photograph of one such system, NGC 2808, is shown in Fig. 1.1. The image appears nearly circular, which is typical for globular clusters accompanying our Galaxy. In more than half these systems the observed

Fig. 1.1. Photograph of NGC 2808. This picture was obtained with the Anglo-Australian Telescope by D. Malin, using an "unsharp masking" technique to bring out both the core and the halo.

[2] ratio of minor axis $2b$ to major axis $2a$ exceeds 0.9; this ratio is generally greater than 0.8. To a first approximation most globular clusters are nearly spherical.

The ages of these systems are determined from their HR diagrams, which show virtually no main-sequence stars with spectra earlier than the "turn-off" point, at which the stars leave the main sequence and evolve for the first time up the giant branch. As time goes on, the position of the turn-off, at which the hydrogen in the stellar core has been converted to helium, moves to stars of lower mass and later spectral type. Thus in principle the position of this point on the observed HR diagram of a cluster can be used to determine the time since the stars, and presumably the entire cluster, were formed. In practice the age is determined by fitting the points in each HR diagram with one of a series of curves, computed with detailed stellar models and with different assumed ages. Some determinations have shown a difference of age between clusters with different abundances of heavy elements with respect to hydrogen (i.e., different metallicities). However, more recent work [3] gives ages between 15 and 18 times 10^9 years for systems with widely different metallicities. Evidently these clusters, in common with other objects of Baade's Population Type II, are generally very old and were likely born in the early phases of galaxy formation. Individual clusters share the high random space velocities generally found for old (Population II) objects.

As indicated in Fig. 1.1, a globular cluster generally shows a large range in surface brightness with increasing radius r. A plot of the observed brightness distribution [4] for this same cluster NGC 2808 is given in Fig. 1.2a, showing a surface brightness, $\sigma(r)$, decreasing by more than four orders of magnitude from the central region, or core, to the outer envelope, or halo. Two quantities used to parametrize such a distribution are the core radius, r_c, defined as the value of r at which the surface brightness is half its central value, and the tidal radius r_t, at which $\sigma(r)$ reaches zero. Determination of r_t requires an extrapolation of the observed points, and depends on the model used for fitting the data. For five-sixths of the clusters measured [5], r_c lies between 0.3 and 10 pc, with r_t/r_c between 10 and 100.

While most systems show a central region or core within which $\sigma(r)$ changes slowly, as indicated in Fig. 1.1, in some systems $\sigma(r)$ increases down to the smallest radius resolved by the seeing. Such systems are said to possess central "cusps," as shown in Fig. 1.2b by the plot [6] of $\sigma(r)$ for the cluster NGC 6624. This cluster is one of the few which contain a strong X-ray burster source, presumably formed in the compact core of the cluster. More detailed knowledge of such cusps would be very relevant to the topic of core collapse, discussed in chapter 4.

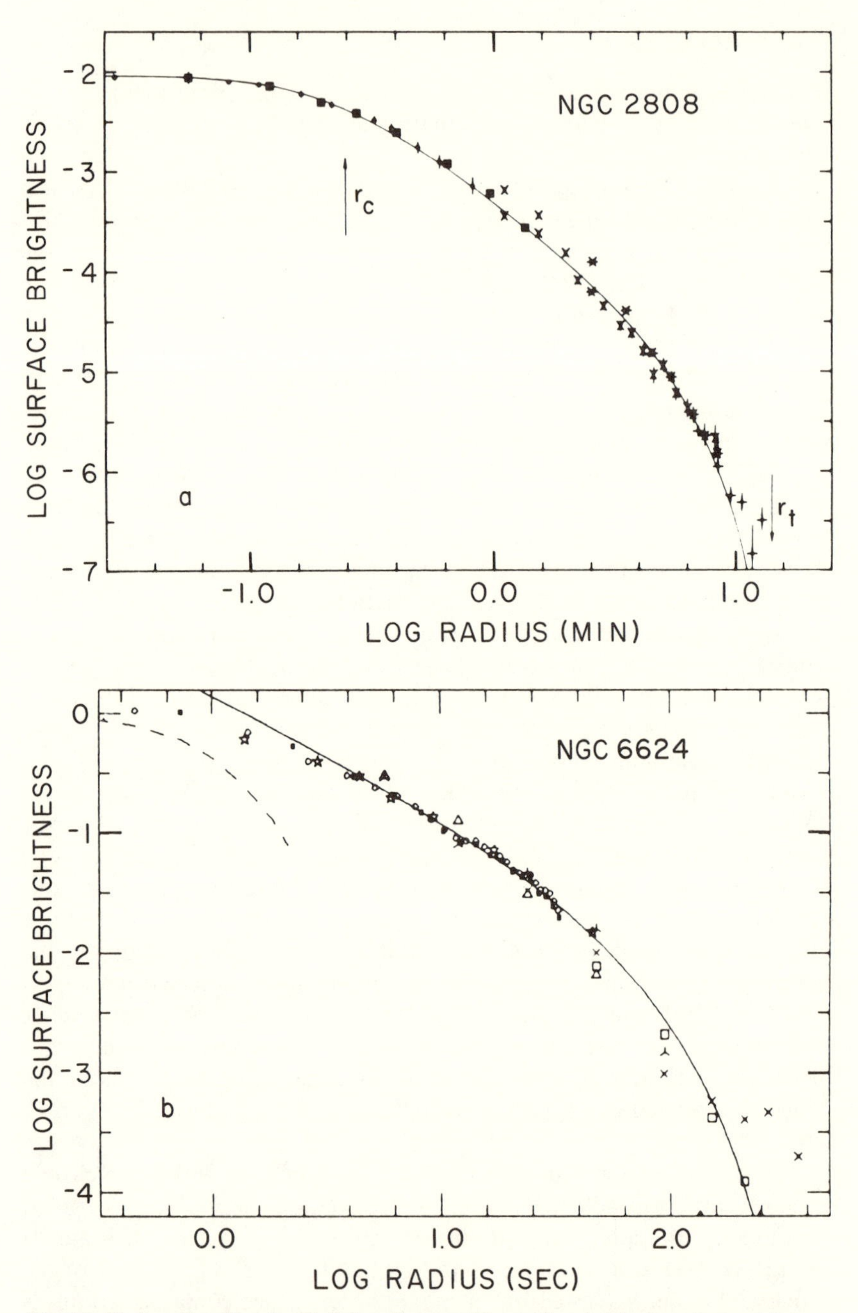

Fig. 1.2. Surface Densities in Globular Clusters. The surface brightness observed in two clusters is plotted against radial distance. Plot a [4] is for NGC 2808; the solid curve shows a theoretical King model with $r_t/r_c = 56$—see §1.2b. Plot b [6] is for NGC 6624, which contains a bright X-ray source within the central core; the solid curve shows a singular King model ($r_c = 0$), while the instrumental profile is indicated by the dashed line. In each curve, different symbols denote results by different observers, with photometry used in the central regions and star counts further out.

Different clusters differ among themselves not only in the distribution of surface density but in other ways as well. Values of the surface brightness at the cluster center [5], denoted by $\sigma(0)$, lie mostly within a range of about 300 to 1. The total visual luminosity L_V varies [1] over a somewhat smaller range, with a dispersion of 0.5 about a mean value of 4.8 for $\log (L_V/L_\odot)$.

To determine accurately the total mass of the cluster requires rather detailed knowledge of the velocity dispersion, which is not easily measured. For M3, the velocity data obtained on about a hundred stars [7] indicate an rms radial velocity of about 5 km/s in the inner regions, decreasing to 2.9 km/s at r equal to 20 pc, some 16 times r_c. These M3 results may be combined with the observed light and stellar luminosity distributions and the best theoretical model, based on equation (1-35) below, to yield a mass of $6.0 \times 10^5\ M_\odot$ and a mass-to-light ratio, M/L_V in solar units, of 2.1. A more approximate method makes use of the central velocity dispersion, obtained from line widths in the integrated spectrum of the cluster center, interpreted with use of the virial theorem; for 10 clusters the values of M/L_V found [8] in this way ranged from about 1 to 4, again in solar units, with masses between 1.7×10^5 and 1.5×10^6 times $M_\odot$. If we assume that the same values of M/L_V apply to the more numerous clusters of lower L_V, for which no velocity data are available, we find that the cluster masses are between 10^4 and $10^6\ M_\odot$ with a peak at about $10^5\ M_\odot$.

One important result of the M3 data is a definite indication of an anisotropic velocity distribution in the cluster halo, with velocities towards or away from the cluster center significantly exceeding the transverse velocities. Indication of a similar conclusion for M3 has been obtained [9] more directly from proper motion data.

With this determination of M/L_V from the velocity observations, together with the measured profiles of $\sigma(r)$, it is possible to compute the particle density of stars, their rms velocity, and finally the value at each point of t_r, the time of relaxation. A precise definition of t_r is postponed until chapter 2, but this quantity may be regarded as the time required for the velocity distribution to approach its Maxwell-Boltzmann form as a result of mutual deflections of stars by each other, accompanied by exchanges of kinetic energy. The time required for dynamical evolution generally exceeds t_r, sometimes by several orders of magnitude. The computed values [5] of $t_r(0) \equiv t_{rc}$, the relaxation time at the cluster center, range mostly from 10^7 up to 10^{10} years.

As we shall see in later chapters, the time interval required for evolution of a cluster as a whole is usually comparable with the value of t_{rh}, the half-mass relaxation time. This reference time is defined—in equation (2-63)—as the value of t_r for average conditions within the radius r_h containing half of the cluster mass. The value of t_{rh} significantly exceeds the central relaxation time. The distribution [10] of estimated t_{rh} values for 32 observed clusters in our Galaxy is shown in Fig. 1.3. Since these values are mostly much less than the cluster ages, there are only a few of these clusters for which relatively little evolution has occurred; many may be highly evolved systems.

One can also compute for each cluster the core dynamical time, t_{dc}, defined as $r_c/v_m(0)$, where $v_m(0)$ is the central rms scalar velocity, equal to $3^{1/2}$ times the dispersion of radial velocities in the core. A typical value of t_{dc} is about 10^5 years, usually less than t_{rc} by a factor 10^{-2} or smaller. In the outer regions of the cluster, for $r \gtrsim r_h$, t_d/t_r is appreciably smaller still.

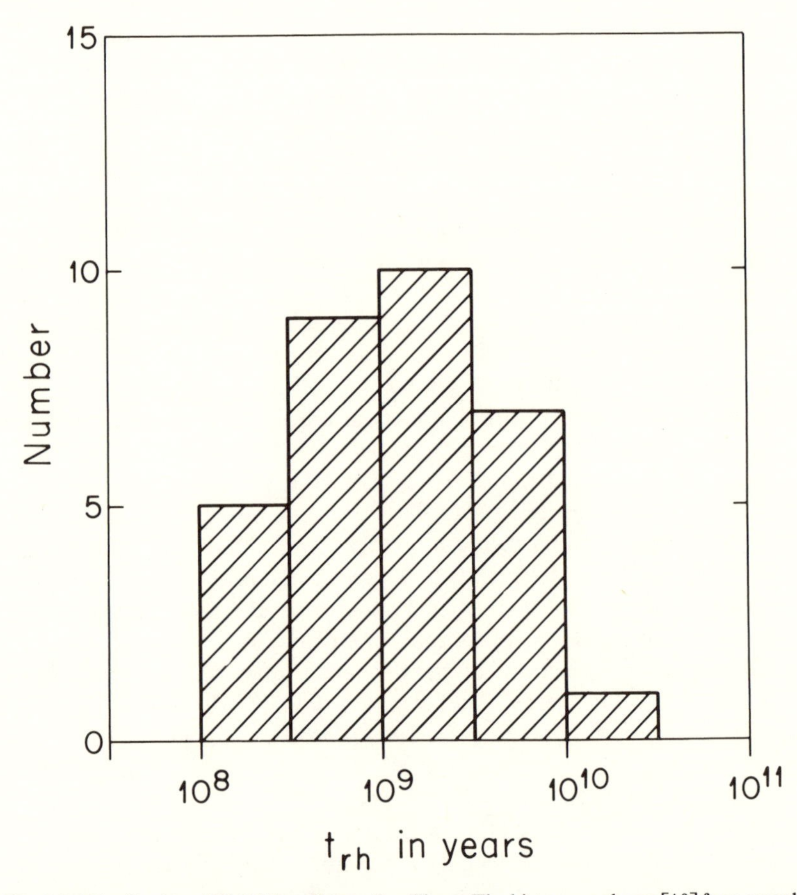

Fig. 1.3. Distribution of Half-Mass Relaxation Times. The histogram shows [10] for a sample of 32 clusters the distribution of values of t_{rh}, the dynamical relaxation time for mean conditions within the inner half of the cluster's mass.

Information on the presence of binary stars in globular clusters is provided [11] by the observed X-ray sources in these systems. Such sources have a bimodal distribution of luminosity, L_x, with bright ones having L_x between 3×10^{35} and 3×10^{37} erg/s, while for a fainter, more numerous group, L_x is less than 10^{34} erg/s; no sources are observed within this gap. The brighter sources are all located within about 1 core radius, r_c, from the cluster centers. The theory of mass stratification, discussed in chapter 3, indicates that the average mass of these systems exceeds the average mass of giant stars in the cluster core (probably about 0.7 $M_\odot$) by a factor between 1.5 and 4.9. If these sources are binaries, the compact object in each of the stronger sources is presumably a neutron star, with a main-sequence dwarf as a companion. The weaker sources are more widely distributed in the clusters where they are observed, consistent with a degenerate dwarf (with a mass somewhat less than that of the giants) as a compact emitting object, and again a dwarf companion which provides the gas for accretion by the X-ray emitting star. The statistics of these two types of objects are consistent with an abundance ratio of roughly 1 to 100 for neutron stars to white dwarfs, though more detailed models would be required for definite results.

Other data do not show many binaries in globular clusters [12], though problems of observational selection complicate the interpretation. For example, of 33 giants observed several times in M3, only one was observed to be a spectroscopic binary [13], although for giant stars in the galactic disc about a third show variations of radial velocity amounting to a few kilometers per second over a few years [7], readily detectable in the M3 data. There appears to be a similar scarcity of spectroscopic binaries among high-velocity (Population II) stars in the galactic disc [14]; in marked contrast, for the more widely separated visual binaries the ratio of single to double stars is independent of stellar velocity. According to the evolutionary theories presented in chapter 7, the X-ray binary stars have been formed in late stages of cluster evolution. The significant correlation between the scarce central cusps and the scarce strong X-ray sources provides strong support for this point of view. The truly primordial binaries present initially in globular clusters may have constituted a relatively small fraction of the total stellar population, though this conclusion is not yet firmly established.

One final observational area where much work has been done on globular clusters is the chemical composition, as determined mostly from spectroscopic studies. The increased average ratio of heavy elements (for example, of Fe) relative to hydrogen with decreasing distance from the galactic center seems firmly established [1]. This result is of great importance for theories of the birth and early evolution of globular clusters in this and other galaxies. However, such theories are still in an early state and are not considered here. More relevant to the evolution of present clusters is the study of chemical inhomogeneities within a single such system [15]. The one cluster in which a marked increase of metallicity all the way into the cluster center seems well established is ω Cen (NGC 5139), for which even in the center the relaxation time has the relatively high value of 0.4×10^{10} years, and scarcely any dynamical evolution may be expected. Hence these results also are more relevant to cluster formation than to evolution. Such effects may exist in some other clusters, especially those with relaxation times longer than average [16], but the evidence is controversial.

1.2 STATIONARY EQUILIBRIA

To facilitate a quantitative discussion of cluster dynamics we introduce idealized models in which the effect of encounters is ignored and steady-state solutions are possible. These models provide a "zero-order" solution for the dynamical equations. Stellar encounters and other effects ignored can then be regarded as perturbations which gradually produce cluster evolution along a sequence of these zero-order models. We first consider the basic assumptions made in the zero-order approximation, together with the basic principles governing their stationary equilibria. Next we review some of the particular zero-order models that have been considered in this connection.

In all these discussions relativistic effects are ignored. The theory of self-gravitating systems with random velocities comparable to the velocity of light may be relevant to galactic nuclei [17], but, as far as we know, not to the globular clusters.

a. Assumptions and principles

The basic simplifying assumption made in the zero-order approximation is:

(A) The granularity of the self-gravitating matter in the cluster can be ignored, and the gravitational potential, taken to be a slowly varying function of position. It is this assumption that makes possible the existence of stationary equilibrium states.

Once this assumption is made, it is useful to define a velocity distribution function $f(\mathbf{r},\mathbf{v},t)$, dependent on the vector position, $\mathbf{r}$, the velocity $\mathbf{v}$, and the time t. This quantity is defined so that $f(\mathbf{r},\mathbf{v},t)\,\mathbf{dr}\,\mathbf{dv}$ is the number of stars at time t within the volume element $\mathbf{dr} \equiv dx\,dy\,dz$ centered at $\mathbf{r}$ and within the velocity space element $\mathbf{dv} \equiv dv_x\,dv_y\,dv_z$ centered at $\mathbf{v}$. This quantity is meaningful if one can construct volume elements which are large enough to contain many stars, but small enough so that conditions are reasonably constant across each element. Under such conditions, $f(\mathbf{r},\mathbf{v},t)$ will be independent of the exact size and shape of the volume element used. Even when the total number of stars, N, in the cluster is as great as 10^6, the instantaneous $f(\mathbf{r},\mathbf{v},t)$ is difficult to define precisely if all six dimensions in $\mathbf{r}$ and $\mathbf{v}$ must be considered. This problem is diminished if, as will be assumed here, some symmetry exists and fewer dimensions enter the argument of $f(\mathbf{r},\mathbf{v},t)$. This difficulty is further eased when conditions in the stellar system are changing relatively slowly, so that in defining the velocity distribution function an average may be taken over many intervals of the dynamical time, t_d, required to cross the cluster, but over a small fraction of the relaxation time, t_r, in which some significant evolution can occur. We have seen above that t_d/t_r is at most about 10^{-2} in the core, and even smaller further out.

The basic relationship which constrains the possible form of f is called the collisionless Boltzmann equation [18], and may be derived from conservation of particles. Relating $\partial f/\partial t$ to the divergence of the particle flows in physical and velocity space gives the familiar result

$$\frac{\partial f}{\partial t} + \sum_i a_i \frac{\partial f}{\partial v_i} + \sum_i v_i \frac{\partial f}{\partial x_i} = 0, \tag{1-1}$$

where x_i denotes one of the three coordinates, x, y or z, a_i is the corresponding particle acceleration (assumed independent of $\mathbf{v}$) and we have denoted $f(\mathbf{r},\mathbf{v},t)$ simply by f. Since $a_i = dv_i/dt$ and $v_i = dx_i/dt$, the left-hand side of equation (1-1) equals Df/Dt, the change of f following a dynamical trajectory; i.e., following a group of stars all with the same v_i and x_i. Evidently the collisionless Boltzmann equation states that the velocity distribution function is constant along such a trajectory.

For almost all the systems we shall study, a_i is the result of a continuous gravitational potential ϕ, and

$$a_i = -\frac{\partial \phi}{\partial x_i}. \tag{1-2}$$

The particle density is given by

$$n(\mathbf{r},t) = \int f(\mathbf{r},\mathbf{v},t)\,\mathbf{dv}, \tag{1-3}$$

where $\mathbf{dv}$ denotes $dv_x\,dv_y\,dv_z$. The mean square velocity, which we denote by v_m^2, is given by

$$n(\mathbf{r},t)v_m^2(\mathbf{r},t) = \int (v_x^2 + v_y^2 + v_z^2) f(\mathbf{r},\mathbf{v},t)\,\mathbf{dv}. \tag{1-4}$$

In equations (1-3) and (1-4) the integrals extend over all velocities.

In principle, equation (1-1) determines the dynamical evolution of a stellar system satisfying assumption (A), subject to Poisson's Law,

$$\nabla^2\phi(\mathbf{r},t) = 4\pi G\rho. \tag{1-5}$$

If the system is a self-gravitating aggregation of stars, ρ is the sum of $n_j m_j$ over all types j, each of number density n_j and mass m_j. A different f_j may be defined for each type, with a subscript j appearing on f, n and v_m in equations (1-1), (1-3) and (1-4). This complication is ignored in the ensuing discussion but can readily be reintroduced.

Equations (1-1) and (1-5) are generally valid and may be applied, for example, to a system evolving on a dynamical time scale. Such a process may have occurred in a collapse phase during the early history of the cluster. In the remainder of this section we shall be interested in the steady-state equilibria of spherical systems. Hence to provide a zero-order approximation, we introduce two additional assumptions.

(B) The gravitational potential ϕ, the velocity distribution function f and all other cluster properties are independent of time.

(C) The cluster has complete spherical symmetry; not only is ϕ a function of r only, but f is a function only of r, v_r and v_t, where $\mathbf{v}_t$ is the velocity transverse to the radius vector $\mathbf{r}$.

Assumption (C) is not necessary in principle for the zero-order solution, but greatly simplifies the models; fortunately this assumption is a realistic one for most clusters except in their outermost regions, where the gravitational field of the Galaxy provides a significant departure from spherical symmetry.

As a result of assumption (B) the energy of a star per unit mass, which we denote by E, is constant, while the angular momentum J per unit mass is constant by virtue of assumption (C). Evidently

$$E = \tfrac{1}{2}v^2 + \phi(r), \tag{1-6}$$

$$J = rv_t = rv \sin\theta, \tag{1-7}$$

where θ is the angle between $\mathbf{r}$ and $\mathbf{v}$. For each E, the value of J cannot exceed $J_c(E)$, the angular momentum of a star in a circular orbit with energy E per unit mass.

The assumed lack of dependence of $f(r,v_r,v_t)$ on the direction of $\mathbf{v}_t$, in the plane perpendicular to $\mathbf{r}$, is an important assumption, which is not a necessary consequence of the spherical symmetry of ϕ or of n. For example, in a system with a completely isotropic distribution of velocities one can imagine reversing the direction of $\mathbf{v}_t$ for all stars revolving in the positive sense about some fixed polar axis; the equilibrium would be unaffected, but the velocity distribution would become strongly anisotropic, with a large net angular momentum about the designated polar axis.

We have seen that f is always constant along a dynamical trajectory, given assumption (A). In a steady state there are, in general, six integrals of motion which are also constant along each dynamical trajectory. It follows directly that f must be some function of these integrals, a result sometimes known as Jeans' Theorem. For motion in the spherically symmetric field envisaged here, the orbit of each star is in a fixed plane and the six integrals are: energy E, angular momentum J, two angles defining the orientation of $\mathbf{J}$ and thus of the orbit plane (orthogonal to $\mathbf{J}$), an angle defining the orientation of the orbit in its plane, and the phase of the star in its orbit at some fixed time. The last of these integrals is irrelevant in a steady state, and the three angles of orientation are irrelevant because of the symmetry assumed for the velocity field. Thus E and J are the only integrals whose variation is significant, and f must be a function of these two quantities. We denote this function by $f(E,J)$, though the mathematical relationship of $f(E,J)$ to the variables E and J is entirely different from that of $f(r,v_r,v_t)$ to its arguments.

An assumed $f(E,J)$ can in principle be readily converted to f as a function of v_r, v_t and r, since from equations (1-6) and (1-7) v_r and v_t can be computed for each "accessible" r for a given E and J provided $\phi(r)$ is known; the accessible values of r will normally lie between a minimum and maximum value, corresponding to the two turning points at pericenter and apocenter, respectively. Also, for each E and r, v_t^2 will have a maximum value, $2[E - \phi(r)]$, reached at a turning point, when v_r is zero; the corresponding value of J at this point is denoted by $J_{\text{max}}(E,r)$. The maximum value of J for a given E, corresponding to the r value in a circular orbit, we have previously denoted by $J_c(E)$.

We give here some of the additional equations satisfied by these steady-state models as a result of their spherical symmetry. Poisson's Law, equation (1-5), becomes

$$\frac{1}{r^2}\frac{d}{dr}\left[r^2\frac{d\phi(r)}{dr}\right] = 4\pi G\rho. \tag{1-8}$$

For a mass component j of the cluster, the Virial Theorem gives

$$M_j v_{mj}^2 = 4\pi \int_0^\infty \rho_j r^3 \frac{d\phi(r)}{dr}\, dr. \tag{1-9}$$

If the system is isolated, with no external fields present, the sum of this equation over all mass components gives the usual relationship

$$2T = -W \approx 0.4GM^2/r_h, \tag{1-10}$$

where T is the total kinetic energy $M\langle v_m^2 \rangle/2$ and W is the total gravitational energy. The total energy, E_T, equals $T + W$ or $W/2$. The quantity r_h in equation (1-10) is the radius containing half the mass; as we shall see below, the corresponding value of r_{hP}, the radius containing half the mass in projection, is generally about $3r_h/4$ for models with an isotropic velocity distribution. For polytropic spheres of index n (see below) the numerical constant in equation (1-10) decreases from 0.44 to 0.38 as n increases from 2 to 5; in the extreme case of a uniform sphere, the constant is 0.48, rising to 0.49 for the Hénon singular model discussed in §3.2. The value of 0.4 adopted should be a reasonable approximation for most systems.

Equation (1-10) states that the internal kinetic energy of an isolated system is half the energy $-W$ required to disassemble the system. However, if stars are removed uniformly from the cluster, the escape energy required per star decreases in proportion to the remaining mass. As a result, the mean energy per unit mass required to remove stars initially, which we write as $\langle v_e^2 \rangle/2$, is twice $-W/M$. Hence we have from equation (1-10)

$$\langle v_e^2 \rangle = 4\langle v_m^2 \rangle. \qquad (1\text{-}11)$$

The brackets here denote averages over the entire cluster mass.

For comparison of theory with observed clusters one must compute the projected mass density $\sigma(r)$, the mass in a column of unit area through the cluster, passing the center at a minimum distance equal to r. For a spherical cluster we obtain

$$\sigma(r) = 2 \int_r^\infty \frac{\rho(r')\,dr'}{[1 - (r/r')^2]^{1/2}}. \qquad (1\text{-}12)$$

The integrated projected mass $\Sigma(r)$ within a circle of radius r is given by

$$\Sigma(\infty) - \Sigma(r) = 2\pi \int_r^\infty r'\sigma(r')\,dr' = 4\pi \int_r^\infty \rho(r')r'^2\,dr'[1 - (r/r')^2]^{1/2}, \qquad (1\text{-}13)$$

as may be verified by differentiation, recovering equation (1-12); since $\Sigma(0)$ vanishes, we see that $\Sigma(\infty)$, the total mass seen in projection, equals M.

b. Analytic models of clusters in equilibrium

A number of models have been obtained based on assumptions (A), (B) and (C), with a variety of analytic assumptions for the velocity distribution function. These models have been useful for theoretical investigations, for initial conditions in the computations of detailed numerical models, and for fits to observed clusters. We discuss here some of these specific models.

The simplest model is a sphere in which $v_r = 0$ and $v_t^2 = GM(r)/r$, just balancing the centrifugal force. The radial density distribution is arbitrary, but if ρ is constant, v_t varies as r and all stars have the same period of revolution about the cluster center, but, of course, a random distribution of orbital planes.

Another class of spherical models is obtained by assuming an isotropic velocity distribution everywhere, and with

$$f = \kappa_1(-E)^p \qquad \text{for } E < 0, \quad \text{and 0 for } E > 0; \tag{1-14}$$

κ_1 is a constant and the potential $\phi(r)$ in equation (1-6) is taken to be zero at the cluster surface. From equation (1-14) we now compute the smoothed density $\rho(r)$, equal to $mn(r)$, where m is the stellar mass. Using equations (1-3) and (1-6) we obtain for $p > -1$.

$$\rho(r) = \kappa_2[-\phi(r)]^{p+3/2} \tag{1-15}$$

where κ_2 is a constant. Equation (1-15) is the basic equation for a polytropic sphere [19], with the usual index n given here by

$$n = p + 3/2. \tag{1-16}$$

Analytic solutions are available for $n = 0$ (a uniform sphere, for which equation (1-14) is inapplicable because of divergence), $n = 1$ and $n = 5$.

Despite its infinite radius, the polytrope with $n = 5$ resembles actual clusters, with a compact core and an extended outer envelope, and has been extensively used as a starting point for numerical model computations. Since it was used by Plummer [20] in an attempt to fit the observed light distributions of clusters, it is often called Plummer's model. Its physical properties are

$$\rho(r) = \frac{3M}{4\pi R^3} \times \frac{1}{[1 + r^2/R^2]^{5/2}}, \tag{1-17}$$

$$M(r) = M \times \frac{r^3/R^3}{[1 + r^2/R^2]^{3/2}}, \tag{1-18}$$

$$\phi(r) = -\frac{GM}{R} \times \frac{1}{[1 + r^2/R^2]^{1/2}} = -2[v_m(r)]^2. \tag{1-19}$$

The scale factor R is related to $\rho(0)$ and to $v_m(0)^2$ through equations (1-17) and (1-19), respectively. From equation (1-12) we find that the projected mass density $\sigma(r)$ is given by

$$\sigma(r) = \frac{4\rho(0)R}{3} \times \frac{1}{(1 + r^2/R^2)^2}. \tag{1-20}$$

The radius r_h containing half the mass equals $1.30R$; in addition, with the use of equation (1-13) one finds that r_{hP}, the radius containing half the projected mass, equals $0.766r_h$ both for Plummer's model and for the uniform sphere.

More realistic clusters are those in which the velocity distribution function approaches its value $f^{(0)}$ in statistical equilibrium, given by

$$f^{(0)} = Ke^{-BE}, \tag{1-21}$$

where K is a constant and

$$B \equiv 3/v_m^2, \tag{1-22}$$

with v_m^2 again denoting the mean square velocity. If equation (1-21) is applicable to the kinetic energy of random motion plus the potential energy, the system is said to be in "kinetic equilibrium." A spherical system in which equation (1-21) holds exactly is called an isothermal sphere. Since such a sphere has infinite mass, it cannot correspond to an actual cluster. However, the inner regions of clusters are believed to be nearly isothermal, and some properties of isothermal spheres are relevant to real clusters. We consider briefly here the properties of such spheres.

Equation (1-8) determines the structure of any of the equilibrium systems considered here. For an isothermal sphere we eliminate ρ with use of equation (1-21) in equation (1-3), substituting for E from equation (1-6). Straightforward integration of equation (1-3), with $4\pi v^2\,dv$ replacing $\mathbf{dv}$, yields

$$\rho(r) = mn(r) = mK\left(\frac{2\pi}{B}\right)^{3/2} e^{-B[\phi(r)]}. \tag{1-23}$$

We now substitute this result into equation (1-8), eliminating B with the use of equation (1-22), and introducing the dimensionless quantities ξ and θ, defined by

$$\xi \equiv \frac{r}{\kappa} \equiv r\left(\frac{12\pi G\rho(0)}{v_m^2}\right)^{1/2}, \tag{1-24}$$

$$\theta(\xi) \equiv B[\phi(r) - \phi(0)]. \tag{1-25}$$

The scale distance κ equals the Jeans length for the values of v_m and ρ at the cluster center. Equation (1-8) now yields

$$\frac{1}{\xi^2}\frac{d}{d\xi}\left(\xi^2\frac{d\theta}{d\xi}\right) = e^{-\theta}. \tag{1-26}$$

Detailed solutions of this equation, subject to the boundary conditions that $\theta = d\theta/d\xi = 0$ at $\xi = 0$, have been computed numerically [21]. For small ξ a series solution can be obtained, and to second order in ξ^2 we have

$$\rho/\rho(0) = e^{-\theta} = 1 - \xi^2/6 + \xi^4/45. \tag{1-27}$$

As ξ increases, θ approaches asymptotically [22], with slow oscillations, the singular solution to equation (1-26).

$$\theta = \ln(\xi^2/2). \tag{1-28}$$

For this simple solution we obtain

$$\rho/\rho(0) = e^{-\theta} = 2/\xi^2 = 2\kappa^2/r^2 \tag{1-29}$$

and

$$M(r) = 8\pi\rho(0)\kappa^2 r. \tag{1-30}$$

For $\xi \geqslant 9$, corresponding to r exceeding about $3r_c$, equations (1-29) and (1-30) give values for the non-singular solution which are correct to within 26 percent or better.

The infinite mass given by equation (1-30), as r increases indefinitely, results from the tail of the Maxwellian velocity distribution in equation (1-21) for large E. If the mass were finite, the energy, E_e, required to escape entirely from the cluster would also be finite, and all stars with a kinetic energy greater than this critical value would escape from the cluster. To maintain equation (1-21) for large E requires incoming as well as outgoing stars at all energies, and these lead directly to the requirement of infinite mass.

A simple velocity distribution function which avoids this infinite mass and which has been successfully used for fitting many observed clusters is the "lowered Maxwellian." In this model the tail of the distribution in kinetic equilibrium is eliminated with the following modified form of equation (1-21),

$$f = \begin{cases} K(e^{-BE} - e^{-BE_e}) & \text{if } E < E_e, \\ 0 & \text{if } E > E_e. \end{cases} \tag{1-31}$$

The physical basis for this distribution function is the presence of a galactic tidal field, which will pull some stars out of the cluster if their distance from the cluster center exceeds r_t, the "tidal cut-off radius."

To obtain a first approximation for this cut-off distance, we regard the Galaxy as a point mass, M_G, situated a fixed distance R_G from the cluster center. Along the line joining the centers of Galaxy and cluster, the tidal force per unit mass, F_t, varies with r according to the relation

$$F_t = \frac{2rGM_G}{R_G^3}. \tag{1-32}$$

This force will be equal and opposite to the gravitational acceleration toward the center of the cluster with mass M_C when $r = r_t$, given by

$$r_t^3 = \frac{M_C}{2M_G} R_G^3. \tag{1-33}$$

A star moving radially along the line of centers will escape from the cluster if it can reach a distance from the cluster center greater than r_t; i.e., if its energy exceeds $\phi(r_t)$. While stars moving in other directions and with some tangential velocities can remain bound with greater energies, a spherically symmetrical model with an isotropic distribution of velocities must contain no stars with E exceeding $E_e = \phi(r_t)$. This objective is accomplished with equation (1-31), which gives a finite cluster mass if r_t is finite. A more detailed discussion of r_t appears in §5.1.

As we shall see in chapter 3, equation (1-31) provides a reasonable approximation for $f(E)$ obtained in some idealized evolving models, taking into account the changes of stellar velocity produced by the granularity of the actual gravitational field.

Models based on a lowered Maxwellian distribution have been computed and compared with observed clusters by King [23], and are generally called King models. The core radius r_c, defined earlier as the value of r at which $\sigma(r)$ is half its central value, is given approximately by

$$r_c = \left[\frac{3v_m(0)^2}{4\pi G\rho(0)}\right]^{1/2} = 3\kappa, \tag{1-34}$$

where κ is the scale distance for the isothermal sphere, given in equation (1-24). Integrations show that by a numerical coincidence equation (1-34) is correct within about half a percent in the limit of large r_t/r_c; however, as r_t/r_c decreases from 100 to 10, $r_c/3\kappa$ decreases from 0.98 to 0.86. At a distance r_c from the center of a King model, the density $\rho(r_c)$ is generally about a third of the central density $\rho(0)$.

In dimensionless form the King models are functions of r_t/r_c only, and for large r_t/r_c are very close to the completely isothermal sphere. The curve used to fit the observations shown in Fig. 1.2a (p. 4) was obtained for $r_t/r_c = 56$. The ratio r_h/r_t varies slowly for these models, rising from 0.12 to 0.21 as r_t/r_c decreases from 100 to 10. The radius r_{hP} containing half the mass in projection equals $0.74r_h$ to within about two percent for this range of r_t/r_c.

A more realistic equation for f takes into account the velocity anisotropy expected under some conditions in the cluster halo. This may be achieved by modifying equation (1-31) in the following manner, somewhat similar to that discussed by Michie [24],

$$f = Ke^{-\gamma J^2}(e^{-BE} - e^{-BE_e}) \qquad \text{if } E < E_e \text{ and } J < J_c(E), \qquad 0 \text{ otherwise.} \tag{1-35}$$

As we have seen earlier, J_c is the value of J for a star in a circular orbit with energy E per unit mass. Michie-King models have been used in detailed fitting of cluster properties [7], with equation (1-35) assumed separately for each of the mass components considered.

1.3 PERTURBATIONS AND THEIR EFFECTS

The steady-state symmetrical models described in the previous section provide a zero-order approximation for the structure of globular clusters. At any one time they can provide an excellent fit for any spherical cluster. The various effects neglected in these models are mostly time dependent and may be regarded as small perturbations of the zero-order model. During the orbital period of a star in the cluster, measured by the dynamical time, t_d, these effects are generally small, but over a period of many orbits they produce gradual changes from one zero-order solution to adjacent ones. Thus these perturbations, which we discuss in the present section, are responsible for the evolution of the cluster.

It is this evolutionary process which presumably accounts for the present state of the clusters that we observe. The zero-order models, of which only a few simple ones were described above, have so much freedom in the choice of f as a function of E and J that they permit an enormous variety of possible clusters. The evolution produced by the various perturbations leads to a state which may in many respects be independent of the initial starting conditions and which, if our theories are correct, may correspond closely to observed clusters.

The perturbations result from the following three main effects: (a) granularity of the gravitational potential within the cluster; (b) gravitational fields produced by masses external to the cluster, especially by the Galaxy; (c) changes in the physical properties of the stars, as a result both of stellar evolution and of direct impact between stars. A qualitative discussion of each of these effects is given in the present section. The quantitative aspects are treated in the following chapters.

a. Granularity of the gravitational field

In a real cluster the gravitational field at any one point will be constantly fluctuating. There will be continuous small changes on the time scale required for the nearest neighboring stars to pass by; this time interval is roughly that required to travel a distance $n^{-1/3}$, about the average distance between stars. There will be slower fluctuations resulting from a slight random excess of stars within regions where the total number of stars is large; these fluctuations have a relatively small amplitude, but produce an appreciable effect because of their longer time scale. Occasionally there will be a transient fluctuation of larger amplitude as a star passes by relatively closely. Finally, there will be occasions when three stars will come close enough together at the same time to affect each other; thus at high densities, three single stars may interact to form a binary system, or a single star can interact with an existing binary. Close encounters involving four or more stars are also possible.

If we postpone for a moment the discussion of binary stars, the main effect produced by all these fluctuations in the gravitational field is to modify the stellar velocities, both in magnitude and direction.

The rate at which stellar velocities are modified by such fluctuations may be calculated by considering the velocity changes produced by a single encounter between two passing stars and summing these changes over all such encounters. Encounters between three single stars are generally quite unimportant in a cluster. Moreover, as we shall see in chapter 2, relatively close encounters between two stars, which produce relatively large changes in the stellar velocities, are generally unimportant; the cumulative effect of many distant encounters, each producing only small changes in velocity, is greater by at least an order of magnitude. Thus the important process is one of random walks in velocity space, with mostly small step sizes occurring. As in other random-walk situations, if the velocity distribution is not characterized by kinetic equilibrium there will be a net diffusion in velocity space, in the direction of reducing the deviations from this equilibrium; i.e., from the Maxwellian distribution, equation (1-21).

This tendency towards a Maxwellian distribution, produced by random two-body encounters, has important consequences for the cluster. The probability of finding a star within an energy interval ΔE at a given energy E is proportional both to exp $(-BE)$—as indicated in equation (1-21)—and to the volume of phase space available for stars within ΔE. Hence one would expect that the tendency toward kinetic equilibrium would promote evolution in two directions—towards more tightly bound gravitational systems, with more strongly negative E, and also towards expansion of the cluster volume, to increase the available phase space. In particular, stars which escape from the cluster entirely have the entire Galaxy to roam around in. These two effects—contraction of the central core and expansion of the outer envelope, with some stars escaping—are complementary. Since the total energy must remain constant, some stars must move to higher energies to absorb the energy given up by the contracting core.

There are three mechanisms by which a cluster composed of single stars evolves in the direction indicated by this statistical argument. The first is the departure of stars whose velocity exceeds the local velocity of escape, a process called evaporation. In an actual cluster, the tail of the Maxwellian distribution, at energies exceeding the escape energy E_e, will be depleted, but the fluctuations in the gravitational field will tend continually to drive some stars up to energies exceeding E_e, and these will quickly escape.

The second mechanism is the tendency towards equipartition among stars of different mass. As a result of this process, the heavier stars will tend to slow down and sink towards the center of the cluster. If the relative number of heavy stars is sufficiently great, this process may run away; the heavy stars, heated by falling inwards under their self-gravitational attraction, may never reach equipartition but will continue to give up to the lighter stars the potential energy released as the heavier ones draw closer together.

The third mechanism, called the gravothermal instability, is a basic process possible in an isothermal sphere, confined within an outer, non-isothermal envelope. An isolated system of self-gravitating masses has negative specific heat, since if it gives up energy it contracts and by the virial theorem increases its kinetic energy. In a confined isothermal sphere, with sufficiently large density contrast between the core and outer regions, the core can collapse and heat up, transferring its energy to the outer regions, which also heat up but by a lesser amount. This collapse accelerates, but the core mass involved also decreases, with the result that for most of the cluster mass the gravothermal collapse is by itself a relatively minor evolutionary episode. However, this phenomenon can produce interesting results within the central regions, including probably the observed X-ray sources, and can lead to an altered structure of the cluster.

Granularity of the cluster potential is produced not only by single stars but also by interactions between stars and binaries and between more complex stellar subsystems, which can strongly affect the course of cluster evolution. Such interactions tend to provide a heat source in the cluster. The binaries can become more tightly bound in an encounter, increasing the translational kinetic energies both of stars and of binaries. Even if the stars gain sufficient energy so that they escape promptly from the cluster, and perhaps the reacting binaries also, this loss of mass by the cluster effectively heats the remaining cluster, exactly as does mass loss by individual stars, discussed below. The three-body encounters which can form binaries are important in a cluster only at very high densities, such as are found during core collapse. Binaries are also formed in important numbers by the process of tidal capture, when two stars pass each other with a minimum separation of only a few radii. As we shall see in chapter 7, the formation of binaries and their subsequent encounters with stars and with each other are believed to play a dominant role during core collapse and the subsequent late evolution of a cluster.

b. External gravitational perturbations

As the cluster moves through space, it will be subject to gravitational forces from various neighboring masses, of which the most important, of course, is the Galaxy. The tidal force which the Galaxy exerts on the cluster has already been discussed above, on the assumption that this force is constant both in magnitude and direction. Such a steady tidal force is of particular importance on the rate of evaporation produced by the fluctuating internal fields within the cluster; a reduction of the energy E_e required to escape from the cluster produces a dramatic increase in the rate at which stars leave the cluster.

An entirely different result follows when the cluster passes through the galactic disc, which most clusters will do twice during each orbital revolution about the Galaxy. During such a passage, which will require only about 10^6 years, the external gravitational field of the galactic disc at the cluster will change dramatically; a uniform field in one direction will change to a field with appreciable divergence, tending to compress the cluster in a direction perpendicular to the galactic plane, and will finally change to a uniform field in the opposite direction. This transient gravitational perturbation will heat the outer parts of the cluster, where the effective external force (i.e., the difference of this force between the cluster center and a region some distance away) is greatest, and where the orbital periods in the cluster exceed the time of passage through the disc. The source of energy for this process is the motion of the cluster as a whole about the Galaxy. The net effect on the cluster is usually to increase its rate of expansion.

A similar but usually somewhat less important effect occurs when a cluster moves in an eccentric orbit around the Galaxy and passes close to the nucleus. The resulting transient tidal force on the cluster can produce appreciable heating of the outermost stars. Most of the cluster will not be much affected, with inner stars changing their orbits adiabatically back and forth as the tidal force varies. Some tidal heating can also be produced when a cluster passes close to a massive cloud in the galactic disc or to another cluster; these processes are usually unimportant compared to heating by passage through the galactic disc or past the nuclear bulge.

c. Physical changes of stars

The chief property of a star which affects cluster dynamics is the stellar mass, which determines the mutual attraction between stars. This mass can change when a star ejects gas. The material ejected generally escapes from a globular cluster; the gas either flows away directly or is swept out by the external medium, with its high ram pressure. Thus stellar mass loss results in a reduction in the total mass, M, of the cluster, decreasing the gravitational binding energy, and leading to an expansion of the cluster. In terms of T and W appearing in the Virial Theorem, T/M is not changed directly by stellar mass loss, since the kinetic energy of all stars per unit mass is unaffected. However, $|W|/M$ decreases as M, with the required energy coming from the kinetic energy of mass ejection by a star and from any subsequent heating of the gas by radiation. Hence after such mass loss $2T$ exceeds $|W|$, a condition for cluster expansion. Thus the net effect of stellar mass loss is similar to that of a heat source within the cluster. We do not include here flow of gas from one component of a binary system to another, since this material does not escape from the cluster. However, as pointed out above, ejection of stars by interaction with binaries has the same effect as mass loss by stars, since again the mass loss is powered by an additional energy source, not included in the kinetic energy T entering the Virial Theorem.

This result may be contrasted to evaporation—mass loss by escape of single stars. In this case the energy needed for escape comes from the kinetic energy of the remaining stars, and T for these stars decreases by at least as much as $|W|$. It follows that $2T$ decreases by more than does $|W|$, a condition for cluster contraction.

We discuss here two avenues by which a star may eject matter which is then lost by the system. One is through normal evolution of a single star. The second results directly or indirectly from a physical impact between two stars.

A star can eject mass at a variety of stages during its lifetime—by stellar winds and by explosions. Both these effects are strongest for the most massive stars; for stars of solar mass, ejection of gas occurs mostly in the giant phase. If clusters were formed with the same initial mass distribution as is observed for young stars in the galactic disc, their first 10^9 years must have been marked by extensive loss of matter from the most massive stars, some of whose remnants—black holes, neutron stars and white dwarfs—may still be present in clusters. The loss of mass from these stars and from their clusters certainly had a significant effect on early cluster evolution. Any heavy compact objects retained from this initial period will concentrate in the cluster core and will have major effects on the structure and evolution of the system.

At present, clusters lack the massive stars which eject material rapidly. The overall mass loss is governed by the rate at which main-sequence stars pass the turn-off point on the Hertzsprung-Russell diagram and travel around this diagram, losing mass and finally ending up as degenerate stars, which gradually fade. The resulting mass-loss rate has been estimated [25] for eight globular clusters from their observed numbers of horizontal-branch stars; a mass loss of $0.2\ M_\odot$ from each such star during a lifetime of 10^8 years was assumed. For each cluster the total mass can be determined from its luminosity, with M/L_V equal to 2.1 in solar units, the value found [7] in M3. The relative loss rate, $\Delta M/M$, during 2×10^{10} years then averages 0.027 with a dispersion of 0.012. Even if the escaping gas has a binding energy substantially above the average in the cluster potential, the heating due to escape of mass at this low rate seems generally of marginal importance. Hence mass loss by stars is not much discussed in this book, which is focussed on present evolutionary processes and ignores problems associated with the formation and early history of the clusters.

A second important mechanism for mass loss and for altering the physical nature of stars is a direct physical impact between two stars, leading to the formation of a single object. Throughout most of a globular cluster's life such impacts are so rare as to be virtually negligible. When a collapsing core produces a large increase in the number of stars per cubic parsec in the central region, such impacts will be more frequent. Their rate is about equal to the rate of tidal capture of one star by another, with the two stars remaining identical. What happens following the impact of two stars is uncertain, though many scenarios have been suggested. One possibility is that after two stars have coalesced, the impact rate with other stars is increased and successive coalescence builds up an object of large mass which very soon undergoes a supernova explosion. If a black hole of appreciable mass remains, this might grow by the capture of other stars.

Another scenario is that the nuclei of the two coalescing stars retain their identities in a common envelope, forming a binary system surrounded by a gaseous blanket. If a single star passes closely by, the binary can react in the same way that more extended binaries do; i.e., the two nuclei can lose energy and approach each other, with the binding energy going into the kinetic energy of separation between the star and the blanketed binary. The separation velocity is likely to be so great that star and binary would both leave the cluster.

Direct impacts between stars must certainly occur during core collapse, and are likely to produce some interesting and important effects. However, in view of the many uncertainties, not much is said in chapter 7 about these various speculative possibilities.

REFERENCES

1. W. E. Harris and R. Racine, *Ann. Rev. Astron. and Astroph.*, **17**, 241, 1979.
2. C. S. Frenk and S. M. Fall, *M. N. Roy. Astron. Soc.*, **199**, 565, 1982.
3. D. A. VandenBerg, *Ap. J. Suppl.*, **51**, 29, 1983.
4. G. Illingworth and W. Illingworth, *Ap. J. Suppl.*, **30**, 227, 1976.
5. R. F. Webbink, *Dynamics of Star Clusters*, IAU Symp. No. 113, ed. J. Goodman and P. Hut (Reidel, Dordrecht), 1985, p. 541.
6. I. R. King, *Dynamics of Star Clusters*, IAU Symp. No. 113, ed. J. Goodman and P. Hut (Reidel, Dordrecht), 1985, p. 1.

7. J. E. Gunn and R. F. Griffin, *A. J.*, **84**, 752, 1979.
8. G. Illingworth, *Ap. J.*, **204**, 73, 1976.
9. K. M. Cudworth, *A. J.*, **84**, 1312, 1979.
10. J. P. Ostriker, L. Spitzer and R. A. Chevalier, *Ap. J.* (*Lett.*), **176**, L51, 1972.
11. J. E. Grindlay, *Dynamics of Star Clusters*, IAU Symp. No. 113, ed. J. Goodman and P. Hut (Reidel, Dordrecht), 1985, p. 43.
12. V. L. Trimble, *Globular Clusters*, IAU Symp. No. 85, ed. J. E. Hesser, (Reidel, Dordrecht), 1980, p. 259.
13. C. P. Pryor, D. W. Latham and M. L. Hazen-Liller, *Dynamics of Star Clusters*, IAU Symp. No. 113, ed. J. Goodman and P. Hut (Reidel, Dordrecht), 1985, p. 99.
14. H. A. Abt, *A. J.*, **84**, 1591, 1979.
15. R. P. Kraft, *Ann. Rev. Astron. and Astroph.*, **17**, 309, 1979.
16. M. S. Chun and K. C. Freeman, *Ap. J.*, **227**, 93, 1979.
17. S. L. Shapiro, *Dynamics of Star Clusters*, IAU Symp. No. 113, ed. J. Goodman and P. Hut (Reidel, Dordrecht), 1985, p. 406.
18. M. Hénon, *Astron. and Astroph.*, **114**, 211, 1982.
19. A. S. Eddington, *Internal Constitution of the Stars* (Cambridge Univ. Press), 1926, chap. 4.
20. H. C. Plummer, *M. N. Roy. Astron. Soc.*, **76**, 107, 1915.
21. S. Chandrasekhar and G. W. Wares, *Ap. J.*, **109**, 551, 1949.
22. S. Chandrasekhar, *An Introduction to the Study of Stellar Structure* (Chicago Univ. Press), 1939, chap. 4, sec. 26.
23. I. R. King, *A. J.*, **71**, 64, 1966.
24. R. W. Michie, *M. N. Roy. Astron. Soc.*, **125**, 127, 1963.
25. R. J. Tayler and P. R. Wood, *M. N. Roy. Astron. Soc.*, **171**, 467, 1975.

2

Velocity Changes Produced by Stellar Encounters

The relative motions of the stars in a globular cluster produce continual fluctuations in the gravitational field, and these fluctuations produce, in turn, changes in the magnitude and direction of each stellar velocity. Thus the energy E of a star and its angular momentum J (both measured per unit mass) will gradually change with time, perturbing the zero-order solutions discussed in the first chapter. In this chapter we discuss the rate at which these changes occur. The stars are treated as point gravitating masses, and hence direct physical collisions, which are generally very infrequent, are ignored. The gravitational interactions considered are all conservative; kinetic energies are exchanged back and forth between stars as a result of their mutual encounters, without any direct change in the total kinetic energy. Moreover, since binary stars and similar bound subsystems are ignored in this chapter, the kinetic energy of the stars resides entirely in the translational energy of their orbital motions through the cluster, and their potential energy results from their gravitational binding energy in the smoothed cluster potential, $\phi(r)$.

The discussion proceeds in three stages. The first section treats changes of velocities produced by encounters between two neighboring stars. The purpose of this discussion is to indicate the relative importance of close and distant encounters and to find in detail how the velocity perturbations vary with local conditions. In the second section, these results are used in deriving the Fokker-Planck equation, which gives the rate of change of the velocity distribution function $f(\mathbf{r},\mathbf{v},t)$. The final equations depend on the properties of adiabatic invariants, discussed in the third section.

2.1 DIFFUSION COEFFICIENTS

To describe quantitatively the rate at which stellar encounters perturb the velocities, we consider $\Delta\mathbf{v}$, the change of velocity of a particular star (which we shall call a "test star") as a result of a single encounter with another passing star. We then sum $\Delta\mathbf{v}$ over all of the encounters experienced by that particular test star per unit interval of time in interactions with all the other stars (which in this treatment are termed "field stars"). This vector sum is denoted by $\langle\Delta\mathbf{v}\rangle$ and, for reasons evident later in this chapter, is called a "diffusion coefficient." Subscripts t and f will be used to denote test and field stars, respectively, though the subscript t will generally be omitted from the diffusion coefficients. Similarly, one can sum $\mathbf{\Delta v \Delta v}$ over all encounters per unit time and determine a tensor coefficient $\langle \mathbf{\Delta v \Delta v} \rangle$. If i, j represent components in some orthogonal coordinates, the diffusion coefficients will have components $\langle \Delta v_i \rangle$ and $\langle \Delta v_i \Delta v_j \rangle$.

a. Relative importance of close and distant encounters

A close encounter between two stars produces relatively large changes in velocity, comparable with the initial velocities. In a distant encounter these changes are smaller. A particular characteristic of inverse-square forces between particles is that the velocity change Δv decreases relatively slowly with increasing distance of closest approach. In consequence, the numerous small velocity changes produced by distant encounters tend to outweigh the effect of close encounters, unlike the normal case of neutral molecules colliding in a gas.

To understand and evaluate the relative importance of encounters at different distances we consider first the change of velocity of a test star in the very simple case where the initial velocity of this star is zero. In addition, the distribution of velocities of the field stars is taken to be isotropic. In this simple situation there is no preferred direction. Hence $\langle \Delta v \rangle$ must vanish, and $\langle (\Delta v_x)^2 \rangle = \langle (\Delta v_y)^2 \rangle = \langle (\Delta v_z)^2 \rangle = \frac{1}{3}\langle (\Delta v)^2 \rangle$. Evidently to second order in Δv there is only one non-zero diffusion coefficient, which we now evaluate.

When two gravitating stars, of masses m_f and m_t, pass by each other, the value of Δv_t resulting from the encounter may be determined from the usual theory for the relative orbit, in which the reduced mass, m_r, equals $m_f m_t / (m_f + m_t)$. Since $J \equiv r^2\, d\theta/dt$ is the angular momentum per unit mass and remains constant in any one orbit, the basic equation for d^2r/dt^2 may be converted into an equation for $d/d\theta(dr/r^2\, d\theta)$, whose solution for r, the distance between the two stars, is the familiar relation

$$\frac{1}{r} = \frac{G(m_f + m)}{J^2}(1 + e \cos \theta), \tag{2-1}$$

where G is the gravitational constant and θ is measured from the direction of minimum r; as in much of the following text, we omit the subscript t from m_t. Equation (2-1) defines a conic section. For the cases of interest here the eccentricity, e, which appears as a constant of integration in equation (2-1), exceeds unity, giving a hyperbola for the relative orbit as shown in Fig. 2.1.

The value of e may be found from the energy integral

$$\left(\frac{dr}{dt}\right)^2 + \frac{J^2}{r^2} - \frac{2G(m_f + m)}{r} = 2E, \tag{2-2}$$

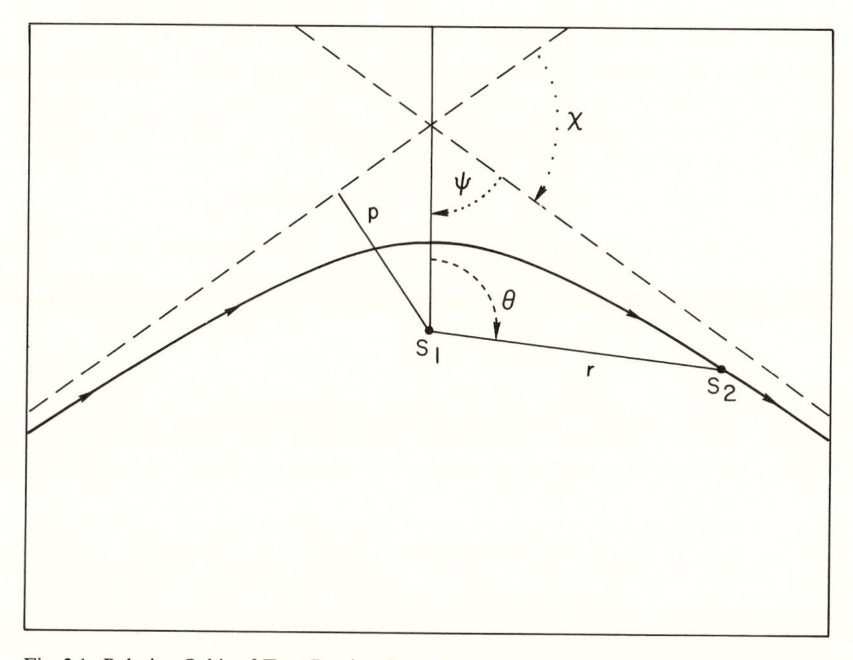

Fig. 2.1. Relative Orbit of Two Passing Stars. The heavy line shows a hyperbolic orbit of star S_2 relative to star S_1. The dashed lines represent the asymptotes. The deflection angle, χ, in their relative orbit is 70°, corresponding to $e = 1.74$.

where E is the total energy divided by the reduced mass. If we again replace dt by $r^2\, d\theta/J$, and substitute for r from equation (2-1), equation (2-2) yields

$$e^2 = 1 + \frac{p^2 V^4}{G^2(m_f + m)^2}, \tag{2-3}$$

where V is the relative velocity for infinite r. In deriving equation (2-3), we have replaced the constant E by $V^2/2$, its value at infinity, and have set the angular momentum J equal to pV, where p is the impact parameter, defined as the distance of closest approach in the absence of collisions.

The angular deflection in the relative orbit is χ, given by

$$\chi = \pi - 2\psi, \tag{2-4}$$

as may be seen in Fig. 2.1. For $\theta = \pi \pm \psi$, r becomes infinite and hence $\cos\psi$ must equal $1/e$. As a result we find from equation (2-3)

$$\tan\psi = \frac{pV^2}{G(m_f + m)} \equiv \frac{p}{p_0}. \tag{2-5}$$

The quantity p_0, defined in equation (2-5), is evidently the impact parameter for which $\psi = \pi/4$ and the deflection χ in the relative orbit equals $\pi/2$.

The motion of the test star relative to the center of gravity of the two stars is equal to the relative velocity V times the factor $m_f/(m_f + m)$. The final relative velocity makes an angle χ with the initial $\mathbf{V}$, and the magnitude of the resulting vector change in $\mathbf{V}$ is $2V \sin \chi/2$. Hence from equation (2-4) we obtain

$$\Delta v_t = \frac{2m_f V}{m_f + m} \sin\left(\frac{\pi}{2} - \psi\right) = \frac{2m_f V \cos \psi}{m_f + m}, \tag{2-6}$$

and with use of equation (2-5) we find, denoting Δv_t by Δv in subsequent equations,

$$(\Delta v)^2 = \frac{4m_f^2 V^2}{(m_f + m)^2} \times \frac{1}{1 + (p/p_0)^2}. \tag{2-7}$$

To calculate $\langle(\Delta v)^2\rangle$ we must multiply equation (2-7) by $2\pi p n_f V\, dp$ and integrate over p from zero to some upper limit p_{max}, which must be taken to be finite since otherwise the integral diverges. Thus we obtain

$$\langle(\Delta v)^2\rangle = \frac{4\pi G^2 n_f m_f^2}{v_f} \ln\left[1 + (p_{max}/p_0)^2\right], \tag{2-8}$$

where we have substituted v_f for V in the final expression, since $v_t = 0$ initially.

The integration leading to equation (2-8) extends over encounters with a wide range in p values. Those with $p < p_0$ are usually called "close encounters" and produce values of the deflection χ exceeding $\pi/2$. Those with large p/p_0 are called "distant encounters"; the value of Δv produced in each such encounter is small, but the cumulative effect of all such encounters provides the dominant contribution to $\langle(\Delta v)^2\rangle$. To obtain typical values for a globular cluster we may take, in accordance with chapter 1, $m = 0.7\ M_\odot$ and $v_f = 8$ km/s, about the three-dimensional rms velocity observed in M3. We then obtain $p_0 = 9 \times 10^{-5}$ pc. Anticipating the discussion below, we set $p_{max} = r_c = 0.3$ pc, a conservatively small value for the core radius. Then $\ln(p_{max}/p_0) = 8$.

We may use this value of $\ln(p_{max}/p_0)$ to evaluate the relative contribution to $\langle(\Delta v)^2\rangle$ of encounters with different ranges of p. In particular, we evaluate $X(p_1)$, defined as the ratio of $\langle(\Delta v)^2\rangle$ resulting from all encounters with $p < p_1$ to the overall value of $\langle(\Delta v)^2\rangle$ in equation (2-8). We find

$$X(p_1) = \frac{\ln\left[1 + (p_1/p_0)^2\right]}{\ln\left[1 + (p_{max}/p_0)^2\right]}. \tag{2-9}$$

Values of $X(p_1)$ are given in Table 2.1 along with the corresponding numbers for χ, the deflection in the relative orbit for $p = p_1$, as computed from equation (2-5).

TABLE 2.1

Relative Importance of Close and Distant Encounters

p_1/p_0	1	2	4	10	60	3×10^3
χ (radians)	1.57	0.93	0.49	0.20	.033	7×10^{-4}
X	0.043	0.10	0.18	0.29	0.51	1.00

We see that the conventional close encounters, with $p \leqslant p_0$, produce only 4 percent of the total $\langle(\Delta v)^2\rangle$. However the contribution by encounters at intermediate distances, producing moderate deflections, is appreciable, with one fifth of $\langle(\Delta v)^2\rangle$ resulting from encounters in which the deflection (in the relative orbit) exceeds 0.40 radians or 23°. These encounters at intermediate distances, producing appreciable deflections, are generally not considered specifically, and the changes of velocity in an actual cluster are generally assumed to result from the cumulative sum of many very small velocity changes. It seems unlikely that the finite velocity changes produced by these encounters at intermediate distances would alter significantly the computed dynamical evolution of clusters, but this conclusion has not been firmly established.

In any case, since p_{max}/p_0 is certainly large, a very good approximation in equation (2-8) is obtained if we set $\ln\left[1 + (p_{max}/p_0)^2\right]$ equal to $2 \ln p_{max}/p_0$. Also we average over the Maxwellian velocity distribution, $f^{(0)}$, obtained from equation (1-21); with K chosen to satisfy equation (1-3) we have

$$f^{(0)} = \frac{j^3}{\pi^{3/2}} n_f e^{-j^2 v^2} \equiv n_f F^{(0)}, \tag{2-10}$$

with

$$j^2 = \frac{1}{2} B = \frac{3}{2v_{mf}^2}, \tag{2-11}$$

where B was introduced in equation (1-21) for $f^{(0)}$ and v_{mf}^2 is the mean square three-dimensional velocity of the field stars. The quantity $F^{(0)}(v_f)$ is the normalized velocity distribution function for the field stars; its integral over $\mathbf{dv}_f$ is unity. Multiplying equation (2-8) by $F^{(0)}(v_f)\,\mathbf{dv}_f$ and integrating, we find the average result

$$\langle(\Delta v)^2\rangle = 3\langle(\Delta v_x)^2\rangle = \frac{4 j n_f \Gamma}{\pi^{1/2}}. \tag{2-12}$$

The quantity Γ is defined by

$$\Gamma \equiv 4\pi G^2 m_f^2 \ln \Lambda, \tag{2-13}$$

with p_{max}/p_0 denoted by Λ. The quantity $\langle(\Delta v_x)^2\rangle$ is the mean square change of test-star velocity in the x direction per unit time; as noted above, for a particle at rest, this is clearly equal to the corresponding quantities in the y and z directions. In the integration we have ignored the variation of $\ln \Lambda$ with v_f, evaluating p_0 at v_{mf}.

For conditions inside the radius r_h, containing half the mass of a cluster, Λ can be evaluated from the Virial Theorem, equation (1-10). If we set $p_{max} = r_h$ and determine p_0 from equation (2-5), with $V^2 = 2v_m^2$, we obtain

$$\ln \Lambda \equiv \ln (p_{max}/p_0) = \ln (0.4N), \tag{2-14}$$

where we have assumed that all stars have the same mass, m_f, and that N, the total number of cluster stars, is accordingly equal to M/m_f. For the most massive clusters N is comparable with 10^6, giving $\ln \Lambda = 13$. For most clusters a value between 10 and 12 is typical; as we have already seen, 8 is a representative value for conditions within the central core, if r_c is as low as 0.3 pc. A more quantitative discussion of $\ln \Lambda$ in the central core appears in §7.1.

It may be noted that equation (2-12) can be obtained equally well from an approximate theory in which all field stars are moving in rectilinear orbits, and the total momentum transmitted, computed by integrating the vector gravitational force over the time, equals that obtained from equation (2-7) with $p/p_0 \gg 1$. In the resultant expression for $\langle(\Delta v_t)^2\rangle$, the integral now diverges at low p/p_0 as well as at high values, and the accurate orbital theory is needed only to provide a value of 1 for the lower limit of p/p_0. Apart from the determination of this minimum p in the evaluation of $\ln \Lambda$, all the velocity changes produced by stellar encounters, as used in most theoretical work, can be computed on the approximation that the perturbing gravitational forces on a test star are those produced by field stars moving in straight lines.

This point of view helps one to understand why stellar encounters have the same effect regardless of whether the impact parameter is small or large compared to the average distance of a star to its nearest neighbor; this mean distance is about equal to $n^{-1/3}$, where n is the number of stars per unit volume. In a globular cluster, $n^{-1/3}$ is typically about 10^{-1} pc in the central region, as compared to a value of about 1 pc for r_c. Thus there will be a great number of field stars at distances between $n^{-1/3}$ and r_c, all of them having encounters at the same time with a particular test star. If all these stars can be assumed to move uniformly along rectilinear paths, the mean square momentum transmitted from each field star to a particular test star can be added to the similar contributions from other stars. More specifically, if the total momentum gained by a test star is summed over time, squared, then averaged over long periods, the only terms differing from zero in the double sum will correspond to $(\Delta v_r)^2$ produced by each passage of a field star, r; i.e., terms $(\mathbf{\Delta v_r}) \cdot (\mathbf{\Delta v_s})$ resulting from the passage of separate stars will cancel out on the average, and the net effect will be the same as if each encounter occurred separately, isolated from the others.

An alternate way of viewing this result is to consider the statistical properties of the overall gravitational field at a point, particularly the auto-correlation function of the gravitational acceleration, $\mathbf{g}(t)$. If $\mathbf{\Delta v}_t$ is the change of velocity experienced by an initially stationary test star at $\mathbf{r} = 0$ during the time interval Δt, then

$$(\Delta v_t)^2 = \iint_0^{\Delta t} \mathbf{g}(t) \cdot \mathbf{g}(t')\, dt\, dt'. \tag{2-15}$$

If Δt is large, the right-hand side equals Δt times the integral over dt of the auto-correlation function of $\mathbf{g}(t)$. If the field stars are assumed to be moving uniformly in straight lines, we can evaluate this function exactly, averaging first over the angle between $\mathbf{r}$ and $\mathbf{v_f}$, then summing over stars at all $r(t)$. For a fixed v_f we obtain

$$\langle \mathbf{g}(t) \cdot \mathbf{g}(t + \tau) \rangle_{\text{Av}} = \frac{4\pi G^2 n_f m_f^2}{v_f} \times \frac{1}{|\tau|}. \tag{2-16}$$

Finally for $\langle (\Delta v)^2 \rangle$ during a unit interval of time equation (2-15) yields

$$\langle (\Delta v)^2 \rangle = \frac{8\pi G^2 n_f m_f^2}{v_f} \ln (\tau_{\max}/\tau_{\min}), \tag{2-17}$$

where $\tau_{\max}$ and $\tau_{\min}$ are arbitrarily chosen to eliminate the divergence of the integral at large and small τ.

Equation (2-17) is identical with equation (2-8), provided $\tau_{\max}/\tau_{\min}$ is set equal to $p_{\max}/p_0$ and this ratio is assumed large. Thus encounters contribute equally to $\langle (\Delta v)^2 \rangle$, regardless of whether the distance of closest approach is much less or much greater than the interparticle distance, $(n_f)^{-1/3}$. Physically, the effect of encounters whose impact parameter exceeds $(n_f)^{-1/3}$ may be viewed as the result of random statistical fluctuations of n_f, producing clumps which extend over substantial distances. These transient clumps in the field-star distribution will have lifetimes comparable to their scale size divided by the random stellar velocity and will produce appreciable acceleration of distant test stars.

There are several considerations which limit $p_{\max}$ in a cluster of stars. If n_f decreases outwards, with a scale distance r_S, the auto-correlation function will be diminished for times greater than r_S/v_m; hence fluctuations of n_f will not contribute appreciably to $\langle (\Delta v)^2 \rangle$ if their scale appreciably exceeds r_S. Another effect is important when diffusion coefficients are computed for test stars in motion, with an orbital period P. Any gravitational perturbation with a time scale much longer than P will produce adiabatic changes in a stellar orbit, with no cumulative change remaining when the perturbation disappears; adiabatic invariants are discussed briefly in §2.3 below. Each of these two effects indicates that for stars at a distance $r > r_c$ from the cluster center p_{max} should be comparable with r, as was assumed in equation (2-14). For $r < r_c$, p_{max} should be comparable with r_c—see §7.1.

b. General diffusion coefficients

We consider next the values of the diffusion coefficients when the test star has a velocity, $\mathbf{v}_t$, in the frame of reference where the field stars have no systematic average velocity. If the velocity distribution of the field stars is again taken to be isotropic, the number of such separate diffusion coefficients is now increased to three, instead of the one discussed above for $v_t = 0$. If we take the x axis to be parallel to $\mathbf{v}_t$, then $\langle(\Delta v_x)^2\rangle$ can now differ from $\langle(\Delta v_y)^2\rangle$ and $\langle(\Delta v_z)^2\rangle$, although these last two quantities must be equal because of symmetry. We denote $\langle(\Delta v_x)^2\rangle$ by $\langle(\Delta v_{||})^2\rangle$, and the sum of $\langle(\Delta v_y)^2\rangle$ and $\langle(\Delta v_z)^2\rangle$ by $\langle(\Delta v_\perp)^2\rangle$. In addition $\langle\Delta v_x\rangle \equiv \langle\Delta v_{||}\rangle$ can differ from zero.

Before deriving the equations for these coefficients when both $\mathbf{v}_t$ and $\mathbf{v}_f$ are finite, we first give a simple evaluation of these three coefficients in the special circumstance that $v_f = 0$, with $\mathbf{v}_t$ finite. This analysis is a useful step in the more general derivation and provides a helpful understanding of how these coefficients behave when v_t much exceeds v_f.

The initial relative velocity $\mathbf{V}$ is now the same for all encounters and equals $\mathbf{v}_t$. The vector change between the initial and final velocity, as the relative velocity rotates through an angle χ, can be resolved into components perpendicular and parallel, respectively, to $\mathbf{V}$. With the use of equations (2-4) and (2-5), we obtain as in equation (2-6)

$$\Delta v_\perp = \frac{m_f}{m + m_f} V \sin\chi = 2V\frac{m_f}{m + m_f} \times \frac{p/p_0}{1 + (p/p_0)^2}, \tag{2-18}$$

$$\Delta v_{||} = -\frac{m_f}{m + m_f} V(1 - \cos\chi) = -2V\frac{m_f}{m + m_f} \times \frac{1}{1 + (p/p_0)^2}. \tag{2-19}$$

If we now multiply the appropriate velocity changes by $2\pi p n_f V\,dp$ and integrate over p, we obtain in the same way as before

$$\langle(\Delta v_\perp)^2\rangle = \frac{2n_f\Gamma}{V}, \tag{2-20}$$

$$\langle(\Delta v_{||})\rangle = -\left(1 + \frac{m}{m_f}\right)\frac{n_f\Gamma}{V^2}, \tag{2-21}$$

$$\langle(\Delta v_{\|})^2\rangle = \frac{n_f \Gamma}{V \ln \Lambda}. \tag{2-22}$$

Terms of order $1/\ln \Lambda$ or p_0/p_{max} have been omitted from equation (2-20) and (2-21). However, in equation (2-22) the dominant term (proportional to $\ln \Lambda$) vanishes, and the non-dominant term is given; the $\ln \Lambda$ factor in the denominator cancels that in Γ—see equation (2-13).

The slowing down of a moving test particle, in accordance with the negative value of $\langle \Delta v_{\|} \rangle$ in equation (2-21), is termed "dynamical friction," and the diffusion coefficient $\langle \Delta v_{\|} \rangle$ is sometimes called the coefficient of dynamical friction. In a steady state the decrease of velocity resulting from dynamical friction is offset, on the average, by the systematic increase of velocity resulting from random walk in velocity space, in accordance with the diffusion coefficients $\langle(\Delta v_{\|})^2\rangle$ and $\langle(\Delta v_{\perp})^2\rangle$. An alternative method of deriving equation (2-21) is to compute $\langle(\Delta v_f)^2\rangle$ produced by the passing test particle, and to require that $m_t v_t \langle \Delta v_{\|} \rangle$, the decrease of kinetic energy of the test particle, just balances the kinetic energy gained by the initially stationary field particles.

Diffusion coefficients of type $\langle(\Delta v)^n\rangle$, where n equals three or more, may also be computed. However, these do not contain $\ln \Lambda$; i.e., their values are determined primarily by relatively close encounters and are generally less important than the dominant terms. While $\langle(\Delta v_{\|})^2\rangle$ has no dominant term in equation (2-22), such a term appears for finite v_f, when the velocity change perpendicular to $\mathbf{V}$ has a component parallel to $\mathbf{v}_t$.

When the field particles have random motions, equations (2-20) and (2-21) still give correctly for each encounter the values of $\langle(\Delta v_{\perp})^2\rangle$ and $\langle \Delta v_{\|} \rangle$ in directions perpendicular and parallel, respectively, to $\mathbf{V}$, the initial relative velocity before an encounter. We ignore $\langle(\Delta v_{\|})^2\rangle$ given by equation (2-22), since this is non-dominant. To make more general use of equations (2-20) and (2-21) we must express these results in terms of a coordinate system with some fixed orientation. Hence, we introduce [1,2] two sets of unit vectors, $\boldsymbol{\varepsilon}$ and $\boldsymbol{\varepsilon}'$. The first set is defined in the inertial reference frame of the stellar system and comprises three unit vectors $\boldsymbol{\varepsilon}_i$ along fixed and mutually orthogonal Cartesian axes; i goes from 1 to 3. The second set comprises another three mutually orthogonal unit vectors $\boldsymbol{\varepsilon}'_k$, with $\boldsymbol{\varepsilon}'_1$ taken to be parallel to $\mathbf{V}$; hence $\boldsymbol{\varepsilon}'_2$ and $\boldsymbol{\varepsilon}'_3$ are perpendicular to $\mathbf{V}$. Evidently the orientation of the $\boldsymbol{\varepsilon}'$ vectors depends on the direction of $\mathbf{V}$.

Equations (2-20) and (2-21) give values of $\langle(\Delta v_{\perp})^2\rangle$ and $\langle \Delta v_{\|} \rangle$ in the $\boldsymbol{\varepsilon}'$ system of axes; we denote values in this reference frame by $\langle(\Delta v_{\perp})^2\rangle'$ and $\langle \Delta v_{\|} \rangle'$. To obtain the velocity changes in the fixed $\boldsymbol{\varepsilon}$ reference frame, we write

$$\langle \Delta v_i \rangle = \boldsymbol{\varepsilon}_i \cdot \boldsymbol{\varepsilon}'_1 \langle \Delta v_{\|} \rangle'. \tag{2-23}$$

For the mean velocity change $\langle \Delta v_i \Delta v_j \rangle$ it is the components of $\boldsymbol{\varepsilon}_i$ and $\boldsymbol{\varepsilon}_j$ along both $\boldsymbol{\varepsilon}_2'$ and $\boldsymbol{\varepsilon}_3'$, each perpendicular to $\mathbf{V}$, that contribute, since it is only $(\Delta v_\perp)^2$ that yields a dominant term. Hence we have

$$\langle \Delta v_i \Delta v_j \rangle = \Xi_{ij} \langle (\Delta v_\perp)^2 \rangle' / 2, \tag{2-24}$$

where by definition

$$\Xi_{ij} \equiv (\boldsymbol{\varepsilon}_i \cdot \boldsymbol{\varepsilon}_2')(\boldsymbol{\varepsilon}_j \cdot \boldsymbol{\varepsilon}_2') + (\boldsymbol{\varepsilon}_i \cdot \boldsymbol{\varepsilon}_3')(\boldsymbol{\varepsilon}_j \cdot \boldsymbol{\varepsilon}_3'). \tag{2-25}$$

The factor 1/2 in equation (2-24) is necessary to give $\langle (\Delta v_k)^2 \rangle'$ separately along each of the two coordinate axes, $k = 2$ and $k = 3$.

In equations (2-20) and (2-21) it was assumed that all field particles had the same initial relative velocity $\mathbf{V}$. In the more general case we let $\mathbf{V} = \mathbf{v} - \mathbf{v}_f$, replace n_f by $f(\mathbf{v}_f)\,\mathbf{dv}_f$ and integrate over $\mathbf{v}_f$. The integral over dp, with V assumed constant, has already been carried out in equations (2-20) and (2-21). The further integration over $\mathbf{dv}_f$ yields

$$\langle \Delta v_i \rangle = -\Gamma(1 + m/m_f) \int \frac{f(\mathbf{v}_f)}{V^2} (\boldsymbol{\varepsilon}_i \cdot \boldsymbol{\varepsilon}_1')\, \mathbf{dv}_f, \tag{2-26}$$

and

$$\langle \Delta v_i \Delta v_j \rangle = \Gamma \int \frac{f(\mathbf{v}_f)}{V} \Xi_{ij}\, \mathbf{dv}_f, \tag{2-27}$$

where as before $\ln \Lambda$ has been evaluated for $V = V_m$, its rms value.

One further pair of simplifications is necessary before these equations are useful. In equation (2-26), $\boldsymbol{\varepsilon}_i \cdot \boldsymbol{\varepsilon}_1'$ equals V_i/V, where V_i is the component of $\mathbf{V}$ along the $\boldsymbol{\varepsilon}_i$ axis. The right-hand side of this equation may be modified so that this factor appears naturally. Since $V^2 = \sum_i V_i^2$, we have $\partial V/\partial V_i = V_i/V$; since also $\partial V_i/\partial v_i = 1$, we may write

$$\frac{\partial}{\partial v_i}\left(\frac{1}{V}\right) = -\frac{1}{V^2}\frac{\partial V}{\partial V_i}\frac{\partial V_i}{\partial v_i} = -\frac{V_i}{V^3} = -\frac{\boldsymbol{\varepsilon}_i \cdot \boldsymbol{\varepsilon}_1'}{V^2}. \tag{2-28}$$

Substituting this result into equation (2-26) gives

$$\langle \Delta v_i \rangle = \Gamma(1 + m/m_f)\frac{\partial h}{\partial v_i} = \Gamma(1 + m/m_f)\frac{v_i}{v}\frac{dh}{dv}, \tag{2-29}$$

where by definition

$$h(v) = \int \frac{f(\mathbf{v}_f)}{|\mathbf{v} - \mathbf{v}_f|}\, \mathbf{dv}_f. \tag{2-30}$$

A similar simplification is possible in equation (2-27) if we make use of the vector identity

$$\boldsymbol{\varepsilon}_i = \sum_{k=1}^{3} [(\boldsymbol{\varepsilon}_i \cdot \boldsymbol{\varepsilon}_k')\boldsymbol{\varepsilon}_k']. \tag{2-31}$$

Evidently any vector can be resolved along some orthogonal set of axes, in this case $\boldsymbol{\varepsilon}'_k$, with components for each k equal to $\boldsymbol{\varepsilon}_i \cdot \boldsymbol{\varepsilon}'_k$. If we now write a similar equation for $\boldsymbol{\varepsilon}_j$ and take the dot product $\boldsymbol{\varepsilon}_i \cdot \boldsymbol{\varepsilon}_j$, this product equals δ_{ij}, since $\boldsymbol{\varepsilon}_i$ and $\boldsymbol{\varepsilon}_j$ are orthogonal unless $i = j$. Similarly only $\boldsymbol{\varepsilon}'_1 \cdot \boldsymbol{\varepsilon}'_1$, $\boldsymbol{\varepsilon}'_2 \cdot \boldsymbol{\varepsilon}'_2$, $\boldsymbol{\varepsilon}'_3 \cdot \boldsymbol{\varepsilon}'_3$ differ from zero and we find

$$\delta_{ij} = (\boldsymbol{\varepsilon}_i \cdot \boldsymbol{\varepsilon}'_1) \cdot (\boldsymbol{\varepsilon}_j \cdot \boldsymbol{\varepsilon}'_1) + (\boldsymbol{\varepsilon}_i \cdot \boldsymbol{\varepsilon}'_2)(\boldsymbol{\varepsilon}_j \cdot \boldsymbol{\varepsilon}'_2) + (\boldsymbol{\varepsilon}_i \cdot \boldsymbol{\varepsilon}'_3)(\boldsymbol{\varepsilon}_j \cdot \boldsymbol{\varepsilon}'_3). \tag{2-32}$$

The last two terms on the right-hand side of this equation are those occurring in equation (2-25) for Ξ_{ij}, which may therefore be written in the form

$$\Xi_{ij} = \delta_{ij} - (\boldsymbol{\varepsilon}_i \cdot \boldsymbol{\varepsilon}'_1)(\boldsymbol{\varepsilon}_j \cdot \boldsymbol{\varepsilon}'_1) = \delta_{ij} - \frac{V_i V_j}{V^2}. \tag{2-33}$$

The right-hand side of this equation may also be obtained if we differentiate V first with respect to v_i, then with respect to v_j, yielding

$$\frac{\partial^2 V}{\partial v_i \, \partial v_j} = \frac{1}{V}\left(\delta_{ij} - \frac{V_i V_j}{V^2}\right) = \frac{\Xi_{ij}}{V}. \tag{2-34}$$

In deriving this result we recall the relationships used in equation (2-28), and also that for $i = j$ there is an additional term $\partial V_i / \partial V_j$ to include. If equation (2-34) is substituted into equation (2-27), this latter equation finally becomes

$$\langle \Delta v_i \Delta v_j \rangle = \Gamma \frac{\partial^2 g}{\partial v_i \, \partial v_j} = \Gamma\left[\frac{\delta_{ij}}{v}\frac{dg}{dv} + \frac{v_i v_j}{v^2}\left(\frac{d^2 g}{dv^2} - \frac{1}{v}\frac{dg}{dv}\right)\right], \tag{2-35}$$

where, again by definition,

$$g(v) = \int f(\mathbf{v}_f) |v - v_f| \, \mathbf{dv}_f. \tag{2-36}$$

and where in the differentiation $\partial/\partial v_i = (v_i/v) \, d/dv$.

It is evident from equation (2-29) that if the i axis is parallel to $\mathbf{v}$, the initial velocity of the test star, then $v_i = v$ and we have

$$\langle \Delta v_{\parallel} \rangle = \Gamma\left(1 + \frac{m}{m_f}\right)\frac{dh}{dv}. \tag{2-37}$$

Examination of equation (2-35) indicates that if $i = j$ and the i axis is again parallel to $\mathbf{v}$, we may write

$$\langle (\Delta v_{\parallel})^2 \rangle = \Gamma \frac{d^2 g}{dv^2}, \tag{2-38}$$

while if $i = j$ and the i axis is perpendicular to $\mathbf{v}$, then $v_i = v_j = 0$, and

$$\frac{1}{2}\langle (\Delta v_{\perp})^2 \rangle = \frac{\Gamma}{v}\frac{dg}{dv}. \tag{2-39}$$

c. Diffusion coefficients for an isotropic velocity distribution

These general results may be applied relatively simply if $f(\mathbf{v}_f)$ is isotropic and hence a function only of the scalar velocity, v_f. In equations (2-30) and (2-36) the integrals over θ, the angle between $\mathbf{v}$ and $\mathbf{v}_f$, are now straightforward, since $f(v_f)$ is independent of θ. If we let $\mu = \cos\theta$, the increment of solid angle is $2\pi\, d\mu$ and the relevant integrals give, after straightforward algebra,

$$2\pi \int_{-1}^{+1} \frac{d\mu}{|v^2 - 2\mu v v_f + v_f^2|^{1/2}} = 4\pi \times \begin{cases} 1/v, & \text{if } v > v_f, \\ 1/v_f, & \text{if } v < v_f; \end{cases} \tag{2-40}$$

$$2\pi \int_{-1}^{+1} |v^2 - 2\mu v v_f + v_f^2|^{1/2}\, d\mu = \frac{4\pi}{3} \times \begin{cases} 3v + v_f^2/v, & \text{if } v > v_f, \\ 3v_f + v^2/v_f, & \text{if } v < v_f. \end{cases} \tag{2-41}$$

Inserting these results in equations (2-30) and (2-36) and integrating over $v_f^2\, dv_f$, we obtain

$$h(v) = 4\pi v\{F_2(v) + E_1(v)\}, \tag{2-42}$$

$$g(v) = \frac{4\pi v^3}{3}\{3F_2(v) + F_4(v) + 3E_3(v) + E_1(v)\}. \tag{2-43}$$

The various functions $E_n(v)$ and $F_n(v)$ are given by

$$F_n(v) \equiv \int_0^v \left(\frac{v_f}{v}\right)^n f(v_f)\, dv_f, \tag{2-44}$$

$$E_n(v) \equiv \int_v^\infty \left(\frac{v_f}{v}\right)^n f(v_f)\, dv_f. \tag{2-45}$$

These results may now be substituted into equations (2-37) through (2-39). With some algebra we obtain

$$\langle \Delta v_{||} \rangle = -4\pi\Gamma\left(1 + \frac{m}{m_f}\right) F_2(v). \tag{2-46}$$

The derivatives with respect to the limits of integration have cancelled out, since the two integrands are equal for $v = v_f$. Similarly we have

$$\langle (\Delta v_{||})^2 \rangle = \frac{8\pi\Gamma v}{3}\{F_4(v) + E_1(v)\}, \tag{2-47}$$

$$\langle (\Delta v_{\perp})^2 \rangle = \frac{8\pi\Gamma v}{3}\{3F_2(v) - F_4(v) + 2E_1(v)\}. \tag{2-48}$$

We evaluate two other useful diffusion coefficients, $\langle \Delta E \rangle$ and $\langle (\Delta E)^2 \rangle$. For the first of these we have

$$\Delta E = v\,\Delta v_{||} + \tfrac{1}{2}(\Delta v_{\perp})^2 + \tfrac{1}{2}(\Delta v_{||})^2, \tag{2-49}$$

which, in view of equations (2-46) through (2-48) gives

$$\langle \Delta E \rangle = 4\pi\Gamma v\left(E_1(v) - \frac{m}{m_f} F_2(v)\right). \tag{2-50}$$

If we square ΔE, only the first term in equation (2-49) gives a dominant contribution, and we have, with use of equation (2-47),

$$\langle (\Delta E)^2 \rangle = v^2 \langle (\Delta v_{||})^2 \rangle = \frac{8\pi\Gamma v^3}{3}\{F_4(v) + E_1(v)\}. \tag{2-51}$$

It can often be assumed that $f(v_f)$ is Maxwellian as well as isotropic and can be taken from equation (2-10). If we let x denote jv_t, where j is defined in equation (2-11), the Maxwellian assumption gives the following standard forms for the diffusion coefficients:

$$\langle \Delta v_{||} \rangle = -2\left(1 + \frac{m}{m_f}\right) n_f \Gamma j^2 G(x), \tag{2-52}$$

$$\langle (\Delta v_{||})^2 \rangle = 2 n_f \Gamma j \frac{G(x)}{x}, \tag{2-53}$$

$$\langle (\Delta v_{\perp})^2 \rangle = 2 n_f \Gamma j \left(\frac{\Phi(x) - G(x)}{x}\right), \tag{2-54}$$

$$\langle \Delta E \rangle = \frac{n_f \Gamma j}{x}\left\{ -\frac{m}{m_f}\Phi(x) + \left(1 + \frac{m}{m_f}\right) x\Phi'(x)\right\}, \tag{2-55}$$

where $\Phi(x)$ is the error function

$$\Phi(x) \equiv \frac{2}{\pi^{1/2}} \int_0^x e^{-y^2}\, dy, \tag{2-56}$$

and where

$$G(x) \equiv \frac{\Phi(x) - x\Phi'(x)}{2x^2}. \tag{2-57}$$

These results can also be derived [3,4] without use of the g and h functions, using the trigonometric transformation of $\mathbf{\Delta v}$ from the $\boldsymbol{\varepsilon}'$ coordinates (defined with respect to the relative orbit) to the $\boldsymbol{\varepsilon}$ frame (defined with respect to the cluster).

Values of the functions $G(x)$ and $\Phi(x) - G(x)$ are given in Table 2.2. For x either large or small these functions have simple expressions. As x

TABLE 2.2
Values of $G(x)$ and $\Phi(x) - G(x)$

x	0.0	0.1	0.2	0.3	0.4	0.5	0.6	0.7	0.8	0.9
$G(x)$	0	.037	.073	.107	.137	.162	.183	.198	.208	.213
$\Phi(x) - G(x)$	0	.075	.149	.221	.292	.358	.421	.480	.534	.584
x	1.0	1.1	1.2	1.3	1.4	1.5	1.6	1.7	1.8	1.9
$G(x)$	.214	.211	.205	.196	.186	.175	.163	.152	.140	.129
$\Phi(x) - G(x)$	.629	.669	.706	.738	.766	.791	.813	.832	.849	.863
x	2.0	2.5	3.0	3.5	4.0	5.0	6.0	7.0	8.0	10.0
$G(x)$	.119	.080	.056	.041	.031	.020	.014	.010	.008	.005
$\Phi(x) - G(x)$	.876	.920	.944	.959	.969	.980	.986	.990	.992	.995

becomes greater than about 3 (i.e., for $v_t \geqslant 2.5v_{fm}$), $\Phi'(x)$ falls below 10^{-4}, $\Phi(x)$ is close to unity and $G(x) = 1/(2x^2)$. In this limit, equations (2-52) and (2-54) yield equations (2-21) and (2-20), respectively, while the dominant term for $\langle(\Delta v_{\|})^2\rangle$ equals $n_f\Gamma/(v_t x^2)$, and for $x^2 > \ln \Lambda$ is less than the non-dominant term in equation (2-22). For small x, series expansions give

$$
\begin{aligned}
G(x) &= \frac{2x}{3\pi^{1/2}}\left(1 - \frac{3}{5}x^2 \ldots\right), \\
\Phi(x) - G(x) &= \frac{4x}{3\pi^{1/2}}\left(1 - \frac{1}{5}x^2 \ldots\right).
\end{aligned}
\qquad (2\text{-}58)
$$

If the constant leading terms in G/x and $(\Phi - G)/x$ are substituted into equations (2-53) and (2-54), respectively, the sum of these two coefficients gives $\langle(\Delta v)^2\rangle$ in equation (2-12).

If the velocity distribution of field particles is not isotropic, additional diffusion coefficients will be non-zero. The general equations (2-29), (2-30), (2-35) and (2-36) remain valid in this situation.

d. Equipartition and relaxation

While the chief use of the diffusion coefficients is to compute how an initial $f(\mathbf{r},\mathbf{v})$ changes with time, these quantities are helpful in other respects also. We use them now to derive the rate of equipartition of energy between a Maxwellian distribution of field stars of mass m_f and a group of test stars which have a different mass, m_t. Both the field stars and the test stars are assumed to have Maxwellian distributions of velocity. The total kinetic energy of a star of each type is defined here as E_{Sf} and E_{St}, in contrast to the previous definition of E as energy per unit mass. The average change of E_{St} per unit time for test stars of velocity v_t is given by equation (2-55), multiplied by m_t. This equation yields a positive energy change for small x and negative for large.

We now assume equation (2-10) for the test stars, denoting by j_t and j_f the value of j for test and field stars, respectively. If we set $j_t = \alpha j_f$, we obtain for $d\bar{E}_{St}/dt$, the change per unit time of $\bar{E}_{St}$, the mean kinetic energy of all the test stars

$$\frac{d\bar{E}_{St}}{dt} = \frac{4}{\pi^{1/2}} \alpha^3 j_f^3 \int_0^\infty v_t^2 e^{-\alpha^2 x^2} dv_t \langle \Delta E_{St} \rangle, \tag{2-59}$$

with x again equal to $j_f v_t$. Interactions among the test stars do not affect $\bar{E}_{St}$, but tend to establish the assumed Maxwellian distribution. With straightforward integration by parts, we obtain

$$\frac{d\bar{E}_{St}}{dt} = 2\left(\frac{6}{\pi}\right)^{1/2} \frac{m_t n_f \Gamma}{m_f} \frac{(\bar{E}_{Sf} - \bar{E}_{St})}{(v_{tm}^2 + v_{fm}^2)^{3/2}}. \tag{2-60}$$

Evidently $\bar{E}_{Sf}$ and $\bar{E}_{St}$ equal $\frac{1}{2} m_f v_{fm}^2$ and $\frac{1}{2} m_t v_{tm}^2$, respectively. The factor $m_t \Gamma / m_f$ is proportional to $m_t m_f$, and $n_t d\bar{E}_{St}/dt + n_f d\bar{E}_{Sf}/dt$ vanishes as required.

Another use of the diffusion coefficients is to determine the "time of relaxation," which measures the time required for deviations from a Maxwellian distribution to be significantly decreased. This time is not, in principle, well defined, since the diffusion coefficients are functions of velocity, and are much less in the high-energy tail of the Maxwellian distribution ($x \gg 1$) than at lower energies. Nevertheless, it is helpful to have some measure of the time required for approaching a Maxwellian distribution; a specific formula for the relaxation time is helpful in comparing different theoretical models as well as different actual clusters.

One definition of the relaxation time which has been used in various theoretical models is defined by the equation

$$t_r \equiv \frac{1}{3} \frac{v_m^2}{\langle (\Delta v_{\|})^2 \rangle_{v=v_m}} = \frac{v_m^3}{1.22 n \Gamma}. \tag{2-61}$$

Thus during the time t_r the mean value of $\Sigma(\Delta v_{\|})^2$ for stars of velocity v_m is equal to $v_m^2/3$, the mean square velocity in any one direction. The numerical value in this equation has been obtained from equation (2-53), with G/x set equal to its value 0.166 for $x = (3/2)^{1/2}$. When a distribution of masses is present, equation (2-61) may still be used [5] if m_f in equation (2-13) is taken as the average mass, and $\frac{1}{2}v_m^2$ is set equal to the average kinetic energy per star, divided by the average mass; however, equation (2-61) becomes somewhat unrealistic for a multi-mass system if this is far from equipartition. If we insert numerical values in equation (2-61), and use practical units, we obtain

$$t_r = \frac{0.065 v_m^3}{n m^2 G^2 \ln \Lambda} = 3.4 \times 10^9 \frac{[v_m(\text{km/s})]^3}{n(\text{pc}^{-3})[m/M_\odot]^2 \ln \Lambda} \text{ years.} \quad (2\text{-}62)$$

A slightly different relaxation time, t_R, is introduced in equation (2-75) in the subsequent section.

For a cluster as a whole, a useful reference time is the half-mass relaxation time, defined as the value of t_r for n equal to the average particle density n_h within the radius r_h, containing half the mass, and for v_m^2 equal to the mean square velocity for the entire cluster, obtained from the virial theorem in equation (1-10). Using equations (2-13) and (2-61), with ln Λ now set equal to 12, we find

$$t_{rh} = 0.138 \frac{N^{1/2} r_h^{3/2}}{m^{1/2} G^{1/2} \ln \Lambda} = \frac{1.7 \times 10^5 [r_h(\text{pc})]^{3/2} N^{1/2}}{[m/M_\odot]^{1/2}} \text{ years.} \quad (2\text{-}63)$$

The half-mass relaxation time has the advantage that it changes relatively little during the evolution of some clusters. By contrast, the central relaxation time $t_r(0)$ decreases markedly during evolution, since the central density $n(0)$ increases much more rapidly than v_m^3 at $r = 0$.

2.2 FOKKER-PLANCK EQUATION

When encounters between particles are important, the change of $f(\mathbf{r},\mathbf{v}t)$ with time is described by the Boltzmann equation, which generalizes equation (1-1) as follows:

$$\frac{Df}{Dt} = \frac{\partial f}{\partial t} + \sum_i a_i \frac{\partial f}{\partial v_i} + \sum_i v_i \frac{\partial f}{\partial x_i} = \left(\frac{\partial f}{\partial t}\right)_{\text{enc}}, \quad (2\text{-}64)$$

where the effect of encounters between particles is included in $(\partial f/\partial t)_{\text{enc}}$. In the full Boltzmann equation this term is expressed generally in terms of f; here we use a simpler presentation, appropriate for encounters which produce small deflections.

We define $\Psi(\mathbf{v},\Delta\mathbf{v})\,\mathbf{d}\Delta\mathbf{v}$ as the probability that during a unit time interval a particle with a velocity $\mathbf{v}$ will experience a velocity change $\Delta\mathbf{v}$ within the range $\mathbf{d}\Delta\mathbf{v}$. Ignoring for the moment the dependence of f on position r, we may write

$$f(\mathbf{v},t+\Delta t) = \int f(\mathbf{v}-\Delta\mathbf{v},t)\Psi(\mathbf{v}-\Delta\mathbf{v},\Delta\mathbf{v})\,\mathbf{d\Delta v}. \tag{2-65}$$

We now expand the left-hand side as a Taylor series in Δt, up to the first-order, replacing Δt by unity, and similarly expand the function $f\Psi$ within the integral as a Taylor series to second order in Δv_x, Δv_y and Δv_z, which we denote by Δv_i, with i from 1 to 3. Equation (2-65) then becomes

$$f(\mathbf{v},t) + \frac{\partial f}{\partial t} = \int \left[f\Psi - \sum_{i=1}^{3} \frac{\partial}{\partial v_i}(f\Psi)\,\Delta v_i + \frac{1}{2}\sum_{i,j=1}^{3} \frac{\partial^2}{\partial v_i\,\partial v_j}(f\Psi)\,\Delta v_i\,\Delta v_j \ldots \right] \mathbf{d\Delta v}, \tag{2-66}$$

where $f = f(\mathbf{v},t)$ and $\Psi = \Psi(\mathbf{v},\Delta\mathbf{v})$. To simplify this expression, we define

$$\langle \Delta v_i \rangle \equiv \int \Psi(\mathbf{v},\Delta\mathbf{v})\,\Delta v_i\,\mathbf{d\Delta v}.$$

$$\langle \Delta v_i \Delta v_j \rangle \equiv \int \Psi(\mathbf{v},\Delta\mathbf{v})\,\Delta v_i\,\Delta v_j\,\mathbf{d\Delta v}. \tag{2-67}$$

Since the integral of $\Psi(\mathbf{v},\Delta\mathbf{v})$ over $\mathbf{d\Delta v}$ in the first term must equal unity, equation (2-66) gives for $\partial f/\partial t$, which we denote by $(\partial f/\partial t)_{\text{enc}}$,

$$\left(\frac{\partial f}{\partial t}\right)_{\text{enc}} = -\sum_{i=1}^{3} \frac{\partial}{\partial v_i}(f\langle \Delta v_i \rangle) + \frac{1}{2}\sum_{i,j=1}^{3} \frac{\partial^2}{\partial v_i\,\partial v_j}(f\langle \Delta v_i\,\Delta v_j \rangle). \tag{2-68}$$

The quantities defined in equations (2-67) are the diffusion coefficients introduced in the previous section. Equation (2-64), with $(\partial f/\partial t)_{\text{enc}}$ substituted from equation (2-68), is called the Fokker-Planck equation.

The third and higher order terms in the Taylor series expansion of $f\Psi$ have been ignored in these equations. This is a valid assumption provided that $\Psi(\mathbf{v},\Delta\mathbf{v})$ becomes very small when $\Delta v/v$ is appreciable. Under this circumstance the higher-order diffusion coefficients are negligible compared to $\langle \Delta v_i \rangle$ and $\langle \Delta v_i\,\Delta v_j \rangle$. Thus the Fokker-Planck equation gives correct results if only the dominant terms, resulting primarily from distant encounters and proportional to $\ln \Lambda$, are included. Non-dominant terms in the diffusion coefficients are sometimes appreciable, but cannot be used in the Fokker-Planck equation, which is invalid for encounters which yield Δv comparable with v.

One might expect that for very small velocity changes the first-order diffusion coefficients should much exceed the quadratic ones—i.e., that $v\langle \Delta\mathbf{v} \rangle$ should greatly exceed $\langle (\Delta v)^2 \rangle$. The fact that these two quantities are of about the same order results from the extensive cancellation of the $\Delta\mathbf{v}$

values produced in successive encounters, an effect which does not occur, of course, when the successive values of $(\Delta v)^2$ are summed.

The Fokker-Planck equation provides a quantitative description of diffusion in velocity space, an effect resulting from small successive velocity changes. The quantity $(\partial f/\partial t)_{\text{enc}}$ in equation (2-68) may be regarded as the divergence of a flux in this space. The first term in this flux results from the systematic velocity drift, $\langle \Delta v_i \rangle$, which as indicated previously is named dynamical friction. The second term is the diffusive flux proper resulting from a random walk in velocity space. This flux is proportional to the gradient of $f\langle (\Delta v)^2 \rangle$, a natural result in any diffusion process.

a. f a function of E only

The Cartesian coordinates in equation (2-68) are not convenient for most applications. We derive now several simplified versions of the Fokker-Planck equation which are valid when f is a function only of energy E and time t, with a similar derivation in the following subsection for the situation in which $f = f(E,J,t)$. While general procedures have been developed [1] for expressing the Fokker-Planck equation in any system of curvilinear coordinates, it is simplest to repeat the derivation of this equation for the situations of interest.

For this purpose, we consider again a spherical system and define $N(E)\,dE$ as the total number of stars with an energy in the interval dE centered at E; we denote by V the volume accessible for such stars. From the definition of $f(\mathbf{r},\mathbf{v},t)$, here replaced by $f(E)$—with the dependence on t usually implied rather than indicated—we write

$$N(E)\,dE = \int dV \int f(E) 4\pi v^2\, dv. \tag{2-69}$$

Since dE equals $v\,dv$ from equation (1-6), this equation reduces to

$$N(E) = 4\pi f(E) \int v\, dV, \tag{2-70}$$

integrated over the spherical volume element $dV = 4\pi r^2\, dr$.

We now compute $(\partial N(E)/\partial t)_{\text{enc}}$ with exactly the same procedure used above for computing $(\partial f/\partial t)_{\text{enc}}$. Thus we define $\Psi(E,\Delta E)\,d\Delta E$ as the probability that during a unit time interval a particle with an energy E will experience an energy change ΔE within the range $d\Delta E$. With use of the same series expansions as before, we obtain

$$\left(\frac{\partial N}{\partial t}\right)_{\text{enc}} = -\frac{\partial}{\partial E}\{N\langle \Delta E\rangle_V\} + \frac{1}{2}\frac{\partial^2}{\partial E^2}\{N\langle (\Delta E)^2\rangle_V\}, \tag{2-71}$$

where the diffusion coefficients $\langle \Delta E\rangle_V$ and $\langle (\Delta E)^2\rangle_V$ are averages over all the stars of energy E within the volume V; the diffusion coefficients are functions of v, and at each value of r, v depends on $\phi(r)$ through equation (1-6). To compute this proper average we integrate $\langle \Delta E \rangle$ and $\langle (\Delta E)^2 \rangle$ over r, weighting each increment dr by its contribution to $N(E)$ in equation (2-70). Hence we have

$$\langle \Delta E \rangle_V = \int \langle \Delta E \rangle v r^2 \, dr \bigg/ \int v r^2 \, dr, \tag{2-72}$$

and similarly for $\langle (\Delta E)^2 \rangle_V$; the integrals in both the numerator and denominator are again taken over the same volume, V, accessible to stars of energy E.

Equation (2-71) may be expressed in two forms which have been used in globular cluster studies. The first is valid for a hypothetical uniform cluster in which the potential ϕ is constant. In this case, v is independent of r for a given E, and the integral in equation (2-70) is simply vV. Since v is a function of E only, $\langle \Delta E \rangle$ and $\langle (\Delta E)^2 \rangle$ are independent of r and can replace $\langle \Delta E \rangle_V$ and $\langle (\Delta E)^2 \rangle_V$ in equation (2-71). If we express $N(E)$ in terms of $f(E)$, with use of equation (2-70), and cancel out the $4\pi V$ factor, equation (2-71) gives $(\partial f/\partial t)_{\text{enc}}$ for a uniform medium with an isotropic velocity distribution. If we use equations (2-50) and (2-51) to express results in terms of the functions E_n and F_n, and replace ∂E by $v\,\partial v$, we find

$$\left(\frac{\partial f}{\partial t}\right)_{\text{enc}} = \frac{4\pi\Gamma}{v^2} \frac{\partial}{\partial v} \left[\frac{m f v^2}{m_f} F_2 + \frac{v^3}{3} \{F_4 + E_1\} \frac{\partial f}{\partial v} \right], \tag{2-73}$$

where f, $\partial f/\partial v$, F_2, F_4 and E_1 are all functions of v (as well as of t, in general). This result is also obtained [1] by transforming equation (2-68) into spherical coordinates.

If we assume that the distribution of field stars is Maxwellian as well as isotropic, equation (2-73) takes the simple form

$$\left(\frac{\partial f}{\partial t}\right)_{\text{enc}} = \frac{1}{t_R} \times \frac{1}{x^2} \frac{\partial}{\partial x} \left[2xG(x) \left(2x \frac{m}{m_f} f + \frac{\partial f}{\partial x} \right) \right], \tag{2-74}$$

where $x = jv$ as in equation (2-52) et seq., and the reference time t_R is defined by

$$\frac{1}{t_R} = \frac{n j^3 \Gamma}{2} = \frac{0.92 n \Gamma}{v_{fm}^3}. \tag{2-75}$$

The quantity t_R, introduced to simplify equation (2-74), has been used as a relaxation time in some discussions of conditions in clusters. From equation (2-61) we see that

$$t_R = \tfrac{4}{3} t_r. \tag{2-76}$$

A second form of equation (2-71) is that applicable to a more realistic spherical cluster, in which ϕ varies with r. For any E the accessible volume includes all radii from zero up to $r_{\max}$, the largest accessible r, at which $\phi(r_{\max}) = E$, and the stellar velocity vanishes. If we again express dV as $4\pi r^2\,dr$, equation (2-70) becomes

$$N(E) = 16\pi^2 pf(E), \tag{2-77}$$

where by definition

$$p(E,t) \equiv \int_0^{r_{\max}} vr^2\,dr = \int_0^{r_{\max}} \{2(E - \phi)\}^{1/2} r^2\,dr. \tag{2-78}$$

The quantity p measures the phase space accessible per unit interval of E. Equation (2-71) now yields, with equation (2-77) used to eliminate $N(E)$,

$$p\left(\frac{\partial f}{\partial t}\right)_{\text{enc}} = \frac{\partial}{\partial E}\left\{-pf\langle(\Delta E)\rangle_V + \frac{1}{2}\frac{\partial}{\partial E}\left[pf\langle(\Delta E)^2\rangle_V\right]\right\}. \tag{2-79}$$

Making use of equation (2-78), we may rewrite equation (2-72) in the simpler form

$$\langle \Delta E \rangle_V = \frac{1}{p}\int_0^{r_{\max}} \langle \Delta E \rangle vr^2\,dr, \tag{2-80}$$

and similarly for $\langle(\Delta E)^2\rangle_V$.

When f is changing with time, the terms for Df/Dt in equation (2-64) must also be considered. Instead of integrating this equation over the volume of phase space per unit energy accessible to stars of a given E, we follow a simpler procedure and derive results from the following two assumptions: (a) f is a function of E and t only; (b) for a family of stars all of the same initial E the temporal variation of the potential ϕ will lead to a changing E. If we denote by Df/Dt the rate of change of $f(E,t)$ for such a family, we have

$$\frac{Df}{Dt} = \frac{\partial f}{\partial t} + \frac{\partial f}{\partial E}\frac{dE}{dt} = \left(\frac{\partial f}{\partial t}\right)_{\text{enc}}. \tag{2-81}$$

To apply this result we evaluate dE/dt on the assumption that the time scale for changes in ϕ is much longer than the orbital periods. The discussion of adiabatic invariants in the final section of this chapter confirms the plausible result that under these conditions dE/dt equals the mean value of $\partial\phi/\partial t$, averaged over the accessible volume of phase space. Such a volume average for an isotropic distribution of velocities is given as in equation (2-80) and we may write

$$\frac{dE}{dt} = \left(\frac{\partial \phi}{\partial t}\right)_V = \frac{1}{p}\int_0^{r_{\max}} v\frac{\partial \phi}{\partial t} r^2\,dr = -\frac{1}{p}\frac{\partial q}{\partial t}, \tag{2-82}$$

where q, a function of E and t, is defined by

$$q(E,t) \equiv \frac{1}{3}\int_0^{r_{\max}} v^3 r^2\, dr = \frac{1}{3}\int_0^{r_{\max}} \{2(E-\phi)\}^{3/2} r^2\, dr. \tag{2-83}$$

The quantity p defined in equation (2-78) is given by

$$p = \frac{\partial q}{\partial E}. \tag{2-84}$$

Generally $q(E,t)$ and $p(E,t)$ are denoted here simply by q and p.

The rate of change of q with time for a family of stars all with the same initial E and with an isotropic velocity distribution, we designate as Dq/Dt. If we write for Dq/Dt an equation analogous to that for Df/Dt but with encounters ignored and if we then eliminate dE/dt and $\partial q/\partial E$ with equation (2-82) and (2-84), we find

$$\frac{Dq}{Dt} = \frac{\partial q}{\partial t} + \frac{\partial q}{\partial E}\frac{dE}{dt} = 0. \tag{2-85}$$

Thus q is a constant for such a family of stars in a slowly variable potential $\phi(r)$. As we shall see in §2.3, q may be regarded as an adiabatic invariant, as was assumed implicitly in equation (2-82).

To obtain the overall equation for the evolution of a spherical cluster with an isotropic velocity distribution, we use equations (2-79) and (2-82) to evaluate $(\partial f/\partial t)_{\text{enc}}$ and dE/dt, respectively, in equation (2-81). The two diffusion coefficients are given in equations (2-50) and (2-51), together with (2-44) and (2-45). In these volume-averaged coefficients we take f to be a function of E only; v and v_f are then the only quantities depending on r and the integrals over $r^2\,dr$ can be expressed in terms of the p and q functions. After some detailed algebra, equation (2-81) becomes

$$\frac{\partial q}{\partial E}\frac{\partial f}{\partial t} - \frac{\partial q}{\partial t}\frac{\partial f}{\partial E} = 4\pi\Gamma\frac{\partial}{\partial E}\left[\frac{m}{m_f} f\int_{-\infty}^{E} f_f \frac{\partial q_f}{\partial E_f}\, dE_f + \frac{\partial f}{\partial E}\left\{\int_{-\infty}^{E} f_f q_f\, dE_f + q\int_E^{\infty} f_f\, dE_f\right\}\right], \tag{2-86}$$

where we have used f_f and q_f to indicate that these quantities are functions of the integration variable E_f rather than of E. In general, both $f_f(E_f)$ and $f(E)$ are non-zero only for values of E_f and E between the central potential, $\phi(0)$, and the escape energy, E_e. For systems with stars all of the same mass, $m = m_f$ and $f_f(E) = f(E)$. If a continuous distribution of masses is present, f_f becomes $f(m_f,E,t)$ and an integral of the right-hand side over dm_f (taking into account that Γ includes a factor m_f^2) is necessary [6].

b. f a function of E and J

For the more general case in which f varies with J as well as with E, we define $N(E,J)\,dE\,dJ$ as the total number of cluster stars within the intervals dE and dJ centered at E and J. The volume accessible to stars of a particular E and J is no longer a sphere but instead a spherical shell, with inner and outer radii equal to r_p and r_a, respectively. The minimum radius corresponds to pericenter, while the maximum radius, to apocenter. At these two radii $v_r = 0$ and the kinetic energy is entirely in the transverse velocity, v_t.

In computing $N(E,J)$ we must take into account that at each radius within this spherical shell a given E and J corresponds to both positive and negative values of v_r; in the following discussion we take v_r to be positive and double the computed number of stars.

The relation between $N(E,J)$ and $f(E,J)$ may be written, integrating over the accessible spherical shell,

$$N(E,J)\,dE\,dJ = 2\int_{r_p}^{r_a} f(E,J) \times 4\pi r^2\,dr \times v^2\,dv \times 2\pi \sin\theta\,d\theta, \qquad (2\text{-}87)$$

where, as in equation (1-7), θ is the angle between $\mathbf{v}$ and $\mathbf{r}$, restricted here to values between 0 and $\pi/2$. Since $dJ = rv_r\,d\theta$, where we make use of the fact that $v_r = v\cos\theta$, and $v_t = v\sin\theta$, we obtain

$$\sin\theta\,d\theta = \frac{v_t\,dJ}{rvv_r} = \frac{J\,dJ}{r^2vv_r}. \qquad (2\text{-}88)$$

Since also $dE = v\,dv$, equation (2-87) now yields, on cancellation of the $dE\,dJ$ factors,

$$N(E,J) = 16\pi^2 Jf(E,J)\int_{r_p}^{r_a}\frac{dr}{v_r} \equiv 8\pi^2 Jf(E,J)P(E,J). \qquad (2\text{-}89)$$

Since $v_r = dr/dt$, the quantity $P(E,J)$, which equals twice the integral in equation (2-89), is simply the orbital period.

The Fokker-Planck equation for $N(E,J)$ may be obtained by a variety of methods [1,2]. Here we compute $(\partial N(E,J)/\partial t)_{\rm enc}$ by following the same approach [7] used above for $(\partial N(E)/\partial t)_{\rm enc}$, and find

$$\left(\frac{\partial N}{\partial t}\right)_{\rm enc} = -\frac{\partial}{\partial E}[N\langle\Delta E\rangle_{\rm orb}] - \frac{\partial}{\partial J}[N\langle\Delta J\rangle_{\rm orb}] + \frac{1}{2}\frac{\partial^2}{\partial E^2}[N\langle(\Delta E)^2\rangle_{\rm orb}]$$
$$+ \frac{\partial^2}{\partial E\,\partial J}[N\langle\Delta E\,\Delta J\rangle_{\rm orb}] + \frac{1}{2}\frac{\partial^2}{\partial J^2}[N\langle(\Delta J)^2\rangle_{\rm orb}], \qquad (2\text{-}90)$$

where N denotes $N(E,J)$. As in §2.2a the diffusion coefficients in equation (2-90) are averages over all the stars included in $N(E,J)$. To carry out this average we again integrate each velocity-dependent diffusion coefficient over dr, now weighting each increment dr by the contribution to $N(E,J)$ from that increment in equation (2-89). Thus, for example, we obtain for $\langle \Delta E \rangle_{\mathrm{orb}}$, the average of $\langle \Delta E \rangle$ for all stars of a particular E and J,

$$\langle \Delta E \rangle_{\mathrm{orb}} = \frac{2}{P(E,J)} \int_{r_p}^{r_a} \langle \Delta E \rangle \frac{dr}{v_r}, \tag{2-91}$$

and similarly for the other diffusion coefficients. It may be noted that these mean diffusion coefficients are equal to averages over individual orbits for a particular E and J, weighting each region of the orbit by the time which a star spends in that region.

These orbital-average diffusion coefficients may be expressed, of course, in terms of the velocity distribution of the field stars, using the results of the previous section. In the inner cluster regions, where most stellar encounters occur, the velocity distribution is nearly isotropic and the diffusion coefficients in equation (2-90) may be determined [7,8] as a function of local conditions (particle density $n(r)$ and velocity distribution function $f(v,r)$) in terms of the three diffusion coefficients discussed in the preceding section.

The full Fokker-Planck equation for $f(E,J,t)$, including the effects of a time-dependent ϕ, is given by equation (2-81). While this equation was given for an isotropic velocity distribution, the additional term $(\partial f/\partial J)(dJ/dt)$ required by the J dependence of f is zero, since J is constant in the absence of encounters. However, equation (2-82) must be modified for the anisotropic case; to take account of the difference in accessible phase space between these two cases, dE/dt is now given by the average of $\partial\phi/\partial t$ over an orbit, a result derived below in equation (2-93). The encounter term $(\partial f/\partial t)_{\mathrm{enc}}$ in equation (2-81) may be found for the anisotropic case from equation (2-90), with equation (2-89) used to express $N(E,J)$ in terms of $f(E,J)$.

2.3 ADIABATIC INVARIANTS

We conclude this chapter with a discussion of the adiabatic invariants in a cluster. The essential results cited here have already been used in the assumption relating dE/dt to the average of $\partial\phi/\partial t$ over accessible phase space. Since further results are somewhat qualitative and have not been applied much to cluster evolution, this section is brief, but is included for logical completeness. We ignore here the direct effects of $(\partial f/\partial t)_{\mathrm{enc}}$ but consider the resultant change of $\phi(r)$ in the course of cluster evolution and the effect of such changes on cluster dynamics.

The motion of single stars rather than of stellar families will be treated first. For a point mass moving in some pootential field the action integrals tend to remain constant [9,10] when the potential changes sufficiently slowly and are called adiabatic invariants. For a spherically symmetric potential the important action variable from this standpoint is the radial action integral $Q(E,J)$—generally a function of time—defined as

$$Q(E,J) \equiv 2 \int_{r_p}^{r_a} v_r \, dr. \tag{2-92}$$

If the time scale t_{per} for some transient perturbation $\Delta\phi$ in potential much exceeds the orbital period $P(E,J)$, then $\Delta Q/Q(E,J)$, the relative change in Q for a particular star, will be small. When there are no complications resulting from resonances between comparable and commensurable periods, it has been shown [11] that $\Delta Q/Q$ goes to zero more rapidly then any power of P/t_{per}, consistent with an exponential variation.

Such a variation, with ln ΔQ varying as $-t_{\text{per}}^2/P^2$ is found in §5.2 with use of a perturbation theory, second order in $\Delta\phi/\phi$, applied to an idealized case. While $Q(E,J)$ for any one star remains nearly constant during slow perturbations, the energy E of a star in an orbit can change appreciably; thus for a fixed value of E, $Q(E,J)$ refers to different stars at different times and is not constant in time.

The constancy of $Q(E,J)$ for each orbit has two important consequences. First, if $(\partial f/\partial t)_{\text{enc}} = 0$, $f(Q,J)$ must be constant with time. A second consequence, which was assumed at the end of §2.2b, is that dE/dt is the orbital average of $\partial\phi/\partial t$. This result may be derived [12] by setting $dQ/dt = 0$ in equation (2-92). The differentiation with respect to the limits of integration gives nothing, since $v_r = 0$ at the limits. With v_r^2 set equal to $2(E - \phi) - J^2/r^2$, the time derivative of the integral yields

$$\frac{dE}{dt} = \left\langle \frac{\partial\phi}{\partial t} \right\rangle_{\text{orb}}. \tag{2-93}$$

The orbital average is defined as in equation (2-91) and we have assumed constant J as usual.

Similar results may be obtained for the situation in which f is a function of E and t only. In this case dE/dt is obtained by averaging equation (2-93) over all orbits for a given E, with each orbit populated in proportion to $N(E,J)\,dJ$ given in equation (2-89). Hence, if we use equation (2-91) as a definition of the orbital average, we obtain

$$\frac{dE}{dt} = 2 \int_0^{J_c} \frac{N(E,J)\,dJ}{N(E)P(E,J)} \int_{r_p}^{r_a} \frac{\partial\phi}{\partial t} \frac{dr}{v_r}, \tag{2-94}$$

where $J_c(E)$, the angular momentum in a circular orbit of energy E, is the maximum possible J for a given E; $N(E)$ is the integral of $N(E,J)\,dJ$ from $J = 0$ to $J = J_c$. With $N(E,J)$ substituted from equation (2-89), equation (2-94) becomes

$$\frac{dE}{dt} = \frac{\int_0^{J_c} J\,dJ \int_{r_p}^{r_a} (\partial\phi/\partial t)\,dr/v_r}{\int_0^{J_c} J\,dJ \int_{r_p}^{r_a} dr/v_r}. \tag{2-95}$$

The denominator in this expression is the integral of $JP\,dJ/2$ and is proportional to the total amount of phase space available per interval dE. The double integrals extend over the region of the r,J plane for which $v_r^2 \geqslant 0$ for a given E and may be evaluated by inverting the order of integration. First we integrate J from 0 to J_{max}, the maximum J for a given E and r; i.e., the value of J at which $v_r = 0$ and r equals either r_p or r_a. Second we integrate r from 0 to $r_{\text{max}}(E)$ as in equation (2-78). After some algebra, we find that the denominator equals p, defined in equation (2-78), and equation (2-95) leads directly to equation (2-82) assumed above.

The function q defined in equation (2-83) plays the same role as does $Q(E,J)$ in the more general case. In fact, one may show, again inverting the order of integration in the r,J plane, that

$$q = \frac{1}{2}\int_0^{J_c} Q(E,J)J\,dJ. \tag{2-96}$$

Thus q is a sum of adiabatic invariants, integrated over $J\,dJ$ for all accessible values of J for a family of stars with some particular E. Since J_c is the angular momentum of a circular orbit of energy E, and must remain constant for a star in that orbit as ϕ and E change, the upper limit of integration in equation (2-96) remains constant for a family of orbits with a particular initial E.

In the general case, as ϕ changes, orbits with different J values in a particular family will change their energies in different ways. For example, a slowly increasing central gravitating mass generates anisotropies in an initially isotropic family [13]. In consequence q is no longer constant for this family. As the different stars in the family acquire different energies, the integral in equation (2-96), which refers to stars all of the same E, no longer represents the same initial family. If we assume that for the orbits in a particular family the values of E all change with time in the same way, remaining equal to each other, then it follows that q is an adiabatic invariant, and stays constant for each family, as ϕ and E slowly change. This result would follow if some additional effect, not specifically included in the $(\partial f/\partial t)_{\text{enc}}$ term, could deflect all stars frequently without changing their energy. Without some such effect q is not an adiabatic invariant and f must depend on J as well as on E. Evidently the assumption of an isotropic velocity distribution, which greatly simplifies the analysis of cluster dynamics, has no firm physical foundation when the mean free path much exceeds the cluster dimensions.

REFERENCES

1. M. N. Rosenbluth, W. M. MacDonald and D. L. Judd, *Phys. Rev.*, **107**, 1, 1957.
2. J. Binney and S. Tremaine, *Galactic Dynamics* (Princeton Univ. Press), 1987, chap. 8.
3. S. Chandrasekhar, *Principles of Stellar Dynamics* (Univ. of Chicago Press), 1942.
4. M. Hénon, *Dynamical Structure and Evolution of Stellar Systems*, ed. L. Martinet and M. Mayor (Geneva Observ.), 1973, p. 183.
5. L. Spitzer and M. H. Hart, *Ap. J.*, **164**, 399, 1971.
6. M. Hénon, *Annales d'Astroph.*, **24**, 369, 1961.
7. A. P. Lightman and S. L. Shapiro, *Ap. J.*, **211**, 244, 1977.
8. S. L. Shapiro and A. B. Marchant, *Ap. J.*, **225**, 603, 1978; see Appendix.
9. D. Lynden-Bell, *Dynamical Structure and Evolution of Stellar Systems*, eds: L. Martinet and M. Mayor (Geneva Observ.), 1973, p. 91.
10. J. Binney and S. Tremaine, *Galactic Dynamics* (Princeton Univ. Press), 1987, sec. 3.6.
11. M. Kruskal, *J. of Math. Phys.*, **3**, 806, 1962.
12. H. N. Cohn, *Ap. J.*, **234**, 1036, 1979.
13. P. Young, *Ap. J.*, **242**, 1232, 1980.

3

Evolution of Idealized Models

The simplified models presented here provide an opportunity to explore a number of physical processes which are important in the dynamical evolution of clusters. In a real cluster or in a realistic model several of these processes will usually occur simultaneously. The idealized models discussed in this chapter make it possible to isolate a number of specific processes and to gain some understanding of the particular effects involved.

These four models, each of which considers some process leading to a cataclysmic fate for the cluster, embody quite different methods of analysis. Evaporation from an isolated cluster, considered in §3.1, is modelled with a spherical cluster of uniform density, for which an accurate numerical solution of the Fokker-Planck equation is readily obtained. Section 3.2 considers a more realistic model, which evaluates the rate of evaporation from a tidally limited cluster, with self-consistent radial distributions of density and potential; the Fokker-Planck equation, now considerably more complicated, is again solved numerically. In §3.3 an isothermal fluid sphere, bounded by a rigid non-conducting shell, is considered and its instabilities examined with the aid of thermodynamics and gas mechanics. In §3.4 a cluster composed of two types of stars with different stellar masses is examined and the condition for equilibrium is analyzed; the dynamical consequences of disequilibrium are discussed qualitatively.

3.1. EVAPORATION FROM AN ISOLATED UNIFORM CLUSTER

In chapter 1 we have seen that the mean square velocity of escape from an isolated cluster is four times the mean square random velocity of stars in the cluster—see equation (1-11). To discuss the consequences of this result in the simplest possible circumstances one may consider a sphere of radius R in which the particle density n is uniform. The distribution of random stellar velocities is assumed isotropic. The potential $\phi(r)$ is idealized as a potential well, with $\phi = 0$ for $r > R$, and $\phi = \phi_0 = -\langle v_e^2 \rangle/2$ inside the sphere. Particles will escape from the sphere, or "evaporate" if their velocity satisfies the condition

$$v^2 > \langle v_e^2 \rangle = 4v_m^2. \tag{3-1}$$

One may obtain a very rough estimate of the evaporation rate by assuming that a Maxwellian distribution is established during the relaxation time t_r. Then the probability ξ_e of evaporation during the time interval t_r may be set equal to the fraction of stars for which $v > 2v_m$ in a Maxwellian distribution. From equation (2-11) we see that when $v/v_m = 2, j^2v^2 = 6$, and from equation (2-10) we obtain, letting $x = jv$ as before,

$$\xi_e = \frac{4\pi}{n_f}\int_{2v_m}^{\infty} f^{(0)}(v)v^2\,dv = \frac{4}{\pi^{1/2}}\int_{2.45}^{\infty} e^{-x^2}x^2\,dx = 7.4 \times 10^{-3}. \tag{3-2}$$

Evidently this determination of ξ_e is very crude, but should give a rough order of magnitude.

a. Evolution of cluster

Before we use the Fokker-Planck equation to determine ξ_e accurately, within the framework of this highly idealized model, we first discuss how the cluster will evolve if the rate of mass loss per unit mass is a constant per time interval dt/t_r. With some modifications this discussion applies also to the self-similar solutions discussed in §§3.2 and 3.3; these are homologous solutions in which the radial variations of density, potential and other physical factors remain invariant with time except for time-dependent scale factors. A contracting uniform sphere, in which all quantities remain independent of r for $r < R$, is evidently such a self-similar model.

The rate of mass loss from the cluster may be written in the form

$$\frac{dM}{dt} = -\frac{\xi_e M}{t_r} = -\frac{\xi_e M(0)}{t_r(0)}\left[\frac{R}{R(0)}\right]^{-3/2}\left[\frac{M}{M(0)}\right]^{1/2}, \tag{3-3}$$

where we denote the cluster mass $M(t)$ and radius $R(t)$ simply as M and R. In obtaining this result we have utilized the functional dependence of t_r (which here is independent of r and hence equals t_{rh}) on $N/N(0) = M/M(0)$ and on $r_h/r_h(0) = R/R(0)$ shown in equation (2-63).

To integrate equation (3-3) over time requires another relation between R and M, which is here obtained from the total cluster energy. This energy changes with time if each escaping star carries away to infinite distance a certain kinetic energy per unit mass, which we denote by ζE_m, where E_m is the mean energy per unit mass of the cluster as a whole. Since E_m is negative, ζ will also be negative if excess energy is carried away by the escapers. The rate of change of the total cluster energy E_T, equal to ME_m, is then given by

$$\frac{dE_T}{dt} = \zeta E_m \frac{dM}{dt} = \frac{\zeta E_T}{M}\frac{dM}{dt}. \qquad (3\text{-}4)$$

Since E_T varies as $-M^2/R$, equation (3-4) leads on integration to

$$\frac{R}{R(0)} = \left[\frac{M}{M(0)}\right]^{2-\zeta}. \qquad (3\text{-}5)$$

If now we use equation (3-5) to eliminate $R/R(0)$ from equation (3-3), integration yields

$$\frac{M}{M(0)} = \left[1 - \frac{\xi_e(7-3\zeta)}{2}\frac{t}{t_r(0)}\right]^{2/(7-3\zeta)} \equiv \left[1 - \frac{t}{t_{\rm coll}}\right]^{2/(7-3\zeta)}, \qquad (3\text{-}6)$$

where $t_{\rm coll}$ denotes the "collapse time," the value of t at which M and R vanish.

The variations with time of E_k, the mean kinetic energy per unit mass, and of v_m, $\rho = 3M/(4\pi R^3)$ and t_r are readily shown to be

$$\frac{E_k}{E_k(0)} = \frac{v_m^2}{v_m^2(0)} = \left[\frac{M}{M(0)}\right]^{-(1-\zeta)}, \qquad (3\text{-}7)$$

$$\frac{\rho}{\rho(0)} = \left[\frac{M}{M(0)}\right]^{-(5-3\zeta)}, \qquad (3\text{-}8)$$

$$\frac{t_r}{t_r(0)} = \left[\frac{M}{M(0)}\right]^{1/2}\left[\frac{R}{R(0)}\right]^{3/2} = 1 - \frac{t}{t_{\rm coll}}. \qquad (3\text{-}9)$$

Since ζ is generally less than unity, this model cluster evidently contracts as it loses mass, with density and mean square velocity approaching infinity as M and R go to zero together. The elapsed time $t_{\rm coll}$ from start to catastrophic finish is less than $t_r(0)/\xi_e$, the initial value of $M/(dM/dt)$, since the rates of evaporation and contraction per unit time increase as t_r becomes steadily smaller.

We designate by τ the time interval $t_{\rm coll} - t$ which remains at each moment until the collapse becomes singular, with R vanishing. If we substitute into equation (3-9) the value of $t_{\rm coll}$ found from equation (3-6), we obtain the simple result

$$\frac{\tau}{t_r} \equiv \frac{t_{\rm coll} - t}{t_r} = \frac{2}{7-3\zeta} \times \frac{1}{\xi_e}. \qquad (3\text{-}10)$$

Thus as the collapse proceeds, with t_r becoming shorter and shorter, the time remaining before complete collapse is a constant factor times the instantaneous relaxation time t_r. Equations (3-6) through (3-9) show that $M/M(0)$ and the other cluster parameters vary as powers of τ.

b. Solution of Fokker-Planck equation

While this uniform model is far from reality, a solution of the Fokker-Planck equation in this case gives a value for the evaporation rate ξ_e and provides a simple illustration of some relevant principles. The analysis presented here is a generalization of an earlier version [1], which ignored the detailed evolution of the cluster and hence gave correctly only the initial value of the mass-loss rate.

In this model the stars are assumed to move within the flat-bottomed potential well introduced at the beginning of §3.1. Stars whose velocity is less than $2v_m$ cannot escape directly but diffuse in velocity space according to the Fokker-Planck equation. Any star whose velocity increases above $2v_m$ is assumed to disappear immediately. In fact, stars would surmount the assumed potential barrier only if their radial velocity v_r exceeded $2v_m$, and as a result the velocity distribution would become anisotropic. Such anisotropy is ignored here, and $f(E,J,t)$ is assumed to be independent of J; escape occurs for $E > 0$, and in the limit where $t_{rh} \gg R/v_m$ this escape is nearly instantaneous, giving $f(E,t) = 0$ for $E \geqslant 0$.

Another simplification of this highly idealized problem is the use of equation (2-74) to compute $(\partial f/\partial t)_{\rm enc}$ in equation (2-64). The assumption that the velocity distribution of the field stars is Maxwellian as well as isotropic gives rather accurately the rate at which stars diffuse up to the escape velocity. However, this approximation ignores the loss of energy by the field particles, a loss which must equal the corresponding energy gained by the escaping test stars; this effect is taken into account below by terms on the left-hand side of equation (3-12). In its exact form, with the diffusion coefficients given by equations (2-29) and (2-35), $(\partial f/\partial t)_{\rm enc}$ must be energy conserving, since energy is conserved in each encounter between stars.

The terms in $\partial f/\partial v_i$ and $\partial f/\partial x_i$ in equation (2-64) must be retained, since divergences both in velocity space and physical space are present in this contracting cluster. We evaluate these terms with the results obtained above for the temporal behaviour of the cluster parameters. Since the escaping stars have a negligible excess energy, we set $\zeta = 0$. We assume that for stars with velocity v the acceleration a, in addition to that produced directly by $(\partial f/\partial t)_{\rm enc}$, is given by

$$a = \frac{v}{v_m}\frac{dv_m}{dt}, \tag{3-11}$$

where dv_m/dt is given by equations (3-7) and (3-6). Equation (3-11) gives a self-consistent result when multiplied by v and averaged over velocity. The linear relation assumed between a and v greatly simplifies the analysis and is valid for pure compressional heating (see below). However, other plausible functional relations should give about the same cluster evolution, provided that dE_k/dt, the net rate of increase of the kinetic energy per unit mass, is correctly given. Since the relaxation time for stars with $v \leqslant v_m$ is much shorter than the evolution time, the velocity distribution will become nearly Maxwellian for these stars, regardless of which ones are most rapidly accelerated by effects other than mutual stellar encounters. Hence equation (3-11) should provide an adequate approximation.

To clarify the physical processes which effect the acceleration a and hence $E_k \equiv \frac{1}{2}v_m^2$, we digress momentarily to evaluate the rates of the two processes which contribute to dE_k/dt. During compressional heating of the cluster by the contracting boundary at $r = R$, $v_m R$ remains constant in the absence of other effects, if the velocity distribution is isotropic; since R varies as M^2, according to equation (3-5), this process makes a contribution $-4E_k\, dM/dt$ to $M\, dE_k/dt$. The kinetic energy lost by the field stars in heating up the escapers is $3E_k\, dM/dt$, since on the average each escaper requires an additional energy (per unit mass) of $3E_k$ to achieve the escape kinetic energy of $4E_k$; this energy loss offsets the energy increase associated with $(\partial f/\partial t)_{\text{enc}}$ as a result of the approximation that the velocity distribution of the field stars is Maxwellian. The final net value of $M\, dE_k/dt$ is $-E_k\, dM/dt$, which follows from equation (3-11) and which gives constant total kinetic energy, ME_k, as required by equation (3-7) for $\zeta = 0$.

We can now complete the Fokker-Planck equation by inserting the proper divergence terms in equation (2-64), with equation (3-11) used for a. Since a is a function of v, the term $a_i\, \partial f/\partial v_i$ in equation (2-64) must be replaced by $\partial(a_i f)/\partial v_i$ to give the correct divergence. Similarly there is a term resulting from the convergence of the mean velocity, associated with the contraction; there will be a slight excess of inwards velocities over outwards velocities. On the average this term must equal $f\, dR^3/R^3\, dt$. Combining all these terms, we find that equation (2-64) becomes

$$\frac{\partial f}{\partial t} + \frac{\partial f}{\partial v}\frac{v}{v_m}\frac{dv_m}{dt} + 3f\frac{dv_m}{v_m dt} + f\frac{dR^3}{R^3\, dt} = \left(\frac{\partial f}{\partial t}\right)_{\text{enc}}. \tag{3-12}$$

The third term represents $f \sum_i \partial a_i/\partial v_i = f\partial(v^2 a)/v^2\, \partial v$; the factor 3 results from the three-dimensional convergence in velocity space.

Assuming a self-similar solution for $f(v,t)$, we substitute into equation (3-12) a solution for f of the form

$$f(v,t) = \frac{nj^3}{\pi^{3/2}}\, g(jv), \tag{3-13}$$

where $j(=(1.5)^{1/2}/v_m)$ and $n(=\rho/m)$ vary with time according to equations (3-7) and (3-8). The time derivative of $\pi^{3/2}f$ now gives two terms; the first is $nj^3 g'v\,dj/dt$, the second, $g\,d(nj^3)/dt$, where $g'(x) \equiv dg(x)/dx$. The first of these terms cancels the second term on the left-hand side of equation (3-12), since $dv_m/v_m\,dt = -dj/j\,dt$. The other time-dependent terms on the left-hand side of equation (3-12) all combine, if we interrelate nj^3, v_m and R^3 with use of equations (3-5) through (3-8); if also equations (2-74) and (2-75) are used for $(\partial f/\partial t)_{\text{enc}}$, we obtain

$$-\frac{2}{7}g\frac{d(nj^3)}{nj^3\,dt} = \frac{1}{2}nj^3\Gamma\frac{1}{x^2}\frac{d}{dx}\left[2xG(x)\left\{2x\frac{m}{m_f}g + \frac{dg}{dx}\right\}\right]. \tag{3-14}$$

If we divide through by nj^3, all the time dependent factors combine to give a constant value; i.e., equation (2-61) and (3-9), again with $\zeta = 0$, give

$$\frac{n(0)j^3(0)d(nj^3)}{(nj^3)^2\,dt} = -\frac{d}{dt}\left[\frac{t_r}{t_r(0)}\right] = \frac{1}{t_{\text{coll}}} \equiv \frac{7\xi_e}{2t_r(0)}. \tag{3-15}$$

If now we set $n(0)j^3(0)\Gamma/2$ equal to $1/t_R(0)$, as in equation (2-75), we obtain for equation (3-14), on the assumption that $m = m_f$,

$$-\lambda g = \frac{1}{x^2}\frac{d}{dx}\left[2xG(x)\left\{2xg + \frac{dg}{dx}\right\}\right], \tag{3-16}$$

where we define

$$\lambda \equiv \xi_e t_R(0)/t_r(0). \tag{3-17}$$

Equation (3-16) yields an eigenvalue problem for $g(x)$, subject to the condition that dg/dx vanish at $x = 0$ and that g vanish at $x = 6^{1/2}$. The value of λ for which the first zero of $g(x)$ occurs at $x = 6^{1/2}$ is found numerically [1] to be 1/88. Since $t_R/t_r = 4/3$ from equation (2-76),

$$\xi_e = 8.5 \times 10^{-3}. \tag{3-18}$$

The relatively close agreement between equations (3-2) and (3-18) is clearly somewhat fortuitous.

Equation (3-16) holds also for a hypothetical static environment [1] in which j, R and M are all constant, and f is assumed to vary as $\exp(-\lambda t/t_R)$. The initial mass loss rate per unit time is correctly given in this case, but the time dependence of f is unrealistic.

The eigenfunction corresponding to the value $\lambda = 1/88$ is plotted in Fig. 3.1, where the solid curve shows the ratio of $g(x)$ to the Maxwellian value. Shown by the crosses are corresponding values of this ratio if $g(x)$ is approximated by the lowered Maxwellian distribution in equation (1-31) with $E_e = 0$. The agreement is very close, within a few percent even for very

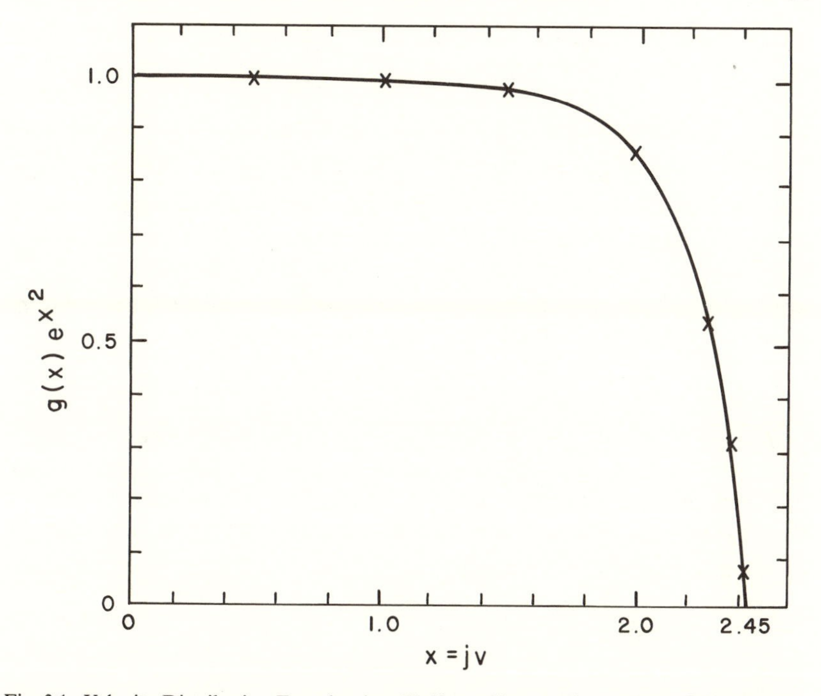

Fig. 3.1. Velocity Distribution Function in a Uniform Cluster. The solid line [1] shows, for a hypothetical, uniform, evaporating cluster, the ratio of the normalized velocity distribution function, $g(x)$, to the Maxwellian function, $\exp(-x^2)$; $x = jv$, where $1.5/j^2$ equals the mean square three-dimensional velocity, v_m^2. The crosses represent a lowered Maxwellian distribution.

small $g(x)$; this result may be established directly [2] by a series expansion of the Fokker-Planck equation in powers of λ.

3.2 EVAPORATION FROM A TIDALLY LIMITED CLUSTER

While an isolated cluster has a certain theoretical appeal, in that spherical symmetry can be fulfilled exactly, real clusters are generally subject to the tidal force of the Galaxy, and stars will tend to escape if their distance from the cluster center exceeds the tidal cut-off distance r_t, given approximately in equation (1-33). If we represent such a cluster with a spherically symmetrical model, as though it were isolated, but assume that any star with $r > r_t$ escapes, the escape energy required is less than from an isolated system, the reduction amounting to GM/r_t. From equation (1-10), $T/M \approx 0.2GM/r_h$, and with $v_e^2 \approx 4v_m^2$ the mean kinetic energy for escape is about equal to $0.8GM/r_h$. Thus if we denote by γ the relative reduction in escape energy, comparing the tidally limited and the isolated clusters, we obtain

$$\gamma = \frac{GM}{r_t} \bigg/ \frac{0.8GM}{r_h} = \frac{5r_h}{4r_t}. \tag{3-19}$$

The ratio r_h/r_t in a cluster can vary appreciably, but as a typical value we may take 0.145, the result obtained in the Hénon tidally limited model discussed below. Then $\gamma = 0.18$. On the simple theory discussed at the beginning of the previous section, ξ_e is an elementary function of $(v_e/v_m)^2$; with this value of γ, $(v_e/v_m)^2$ is reduced from 4 to 3.3, and from equation (3-2) we find that the escape probability per time interval t_{rh} is

$$\xi_e = 2.0 \times 10^{-2}, \tag{3-20}$$

an increase by about a factor three over the value found in equation (3-2) for an isolated static system. While equations (3-20) and (3-2) are clearly very approximate, they indicate that one may expect an appreciably greater evaporation rate from a tidally limited cluster than from one which is isolated.

a. Evolution of cluster

The evolution of a tidally truncated cluster follows a quite different course from that of an isolated system. As M decreases, r_t also decreases; equation (1-33) indicates that the mean density of the cluster within the radius r_t equals twice the mean density, ρ_G, obtained from the galactic mass, M_G, divided by the volume of a sphere with radius R_G, the distance from the cluster to the galactic center. During evolution the cluster density remains constant and we have

$$\frac{r_t}{r_t(0)} = \left[\frac{M}{M(0)}\right]^{1/3} \tag{3-21}$$

instead of the variation as $[M/M(0)]^2$ found for $R/R(0)$ in equation (3-5) with $\zeta = 0$. In the present case, the value of ζ does not affect the time dependence of the variables; if we assume that ξ_e is constant, the right-hand side of equation (3-3) is independent of time, thanks to equation (3-21), giving

$$M = M(0) - \int_0^t \frac{\xi_e M(0)}{t_{rh}(0)}\, dt = M(0)\left[1 - \frac{\xi_e t}{t_{rh}(0)}\right]; \tag{3-22}$$

in the definition of ξ_e we have replaced t_r, the relaxation time in the uniform sphere, by t_{rh}, given in equation (2-63).

Evidently M decreases at a constant rate down to $M = 0$, when the cluster disappears. The mean particle density n, which we define as $3M/4\pi R_t^3 m$, remains constant, as we have seen. If we denote by E_m and v_m^2 the average values of these quantities over the mass of the cluster, and assume that the cluster contracts homologously, we obtain

$$\frac{E_m}{E_m(0)} = \frac{v_m^2}{v_m(0)^2} = \left[\frac{M}{M(0)}\right]^{2/3}, \tag{3-23}$$

$$\frac{t_{rh}}{t_{rh}(0)} = \frac{n(0)j(0)^3}{nj^3} = 1 - \frac{\xi_e t}{t_{rh}(0)}. \tag{3-24}$$

We see that equation (3-24) is very similar to equation (3-9); in both models t_{rh} decreases linearly with time.

The assumption of self-similar or homologous contraction is difficult for the cluster to obey without some additional source of energy. As the cluster loses mass, its total energy E_T in equilibrium increases at the following rate

$$\frac{dE_T}{dt} = -\frac{d}{dt}\left(\frac{0.2GM^2}{r_h}\right) = -\frac{1}{3}\frac{GM}{r_h}\frac{dM}{dt}, \tag{3-25}$$

where we have used equation (1-10) for T and W and have eliminated $dr_h/r_h dt$ with the use of equation (3-21), assumed valid for $r_h/r_h(0)$. This required increase of energy is provided in part by the negative potential energy carried away by the evaporating stars. If we let $(dE_T/dt)_e$ represent the total energy gained from this source, we have

$$\left(\frac{dE_T}{dt}\right)_e = -\frac{GM}{r_t}\frac{dM}{dt}. \tag{3-26}$$

The energy supply in equation (3-26) will equal that required by equation (3-25) if $r_h = r_t/3$. In fact the ratio r_t/r_h usually exceeds 3, in which case a self-similar contraction requires an additional flow of energy into the cluster. Physically, as the cluster evaporates and contracts homologously, its binding energy becomes less and less. The binding energy of stars when they cross the escape radius r_t is not large enough to offset this reduction in cluster binding energy, and the total energy of the entire system, cluster plus escapers, increases steadily. Hence some source of energy is required.

b. Solution of Fokker-Planck equation

We consider now a detailed model for this case, based on a self-similar solution of the Fokker-Planck equation, with the structure of the cluster taken into account. While no energy source was specifically assumed at the outset, such a source appeared in the solution, which in the center approached the singular solution, equation (1-29), for an isothermal sphere, modified to give an energy flux from the center. In addition, the velocity distribution function was assumed isotropic, which in the outer regions of the cluster is not likely to be valid, but which presents substantial simplifications in the analysis. Apart from these two restrictions, the Hénon model [3] makes no approximations other than those given in §1.1, together with the basic assumption of distant encounters, discussed in chapter 2.

The basic equations to be solved simultaneously are:

a) Equation (1-3) relating n to the integral of $vf\, dE$.
b) Poisson's equation (1-5) relating $\nabla^2\phi$ to nm.
c) Equation (2-83) for the function q.
d) The Fokker-Planck equation (2-86) for a cluster with an isotropic velocity distribution.

As in §3.1, the physical variables may be assumed proportional to time-dependent scaling factors—essentially those given above—multiplied by the homologous variables, which we denote by f_*, E_*, ρ_* and r_*, whose relationship with each other is assumed to be independent of time and can be determined from a single integration, subject to the condition that f_* vanish for $E_* > \phi_*(r_{t*}) \equiv U_*(r_{t*}) = 0$.

The resultant values of f_*, ρ_* and r_* as functions of E_* and U_* are shown in Fig. 3.2. The radius r_h containing half the mass is found to be $0.145r_t$. For comparison with f_* a dashed line shows the lowered Maxwellian distribution obtained from equation (1-31). While there is general qualitative agreement, the quantitative agreement evident in Fig. 3.1 (p. 57) for the uniform cluster is absent here. Though both curves approach $\exp(-E_*)$ for E_* (in dimensionless units) a large negative number, their ratio at intermediate E_* is as great as 1.4. One reason for this difference is that f_* must fall below the Maxwellian curve for increasing E_* in order to drive the outward energy flux that characterizes this solution. While the Hénon model is similar to a King model in the unit of zero core radius, there is evidently a significant difference between them.

According to equation (3-22) the rate of mass loss is constant with time. The detailed numerical solution gives for the escape probability ξ_e, per unit t_{rh}, the value

$$\xi_e = 4.5 \times 10^{-2}. \qquad (3\text{-}27)$$

This result is some five times greater than that found in equation (3-18) for the uniformly contracting sphere, and about twice the crude estimate in equation (3-20). Since t_{rh} varies as $M/\rho_h^{1/2}$, the mass loss rate per unit time

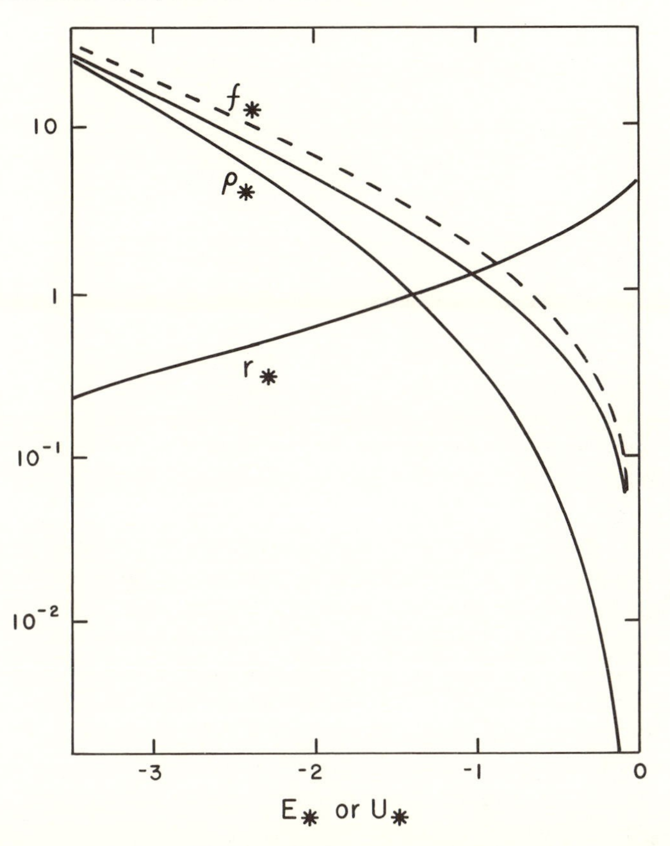

Fig. 3.2. Structure of Hénon Self-Similar Model. The solid lines show the variation through the Hénon model cluster [3] of the homologous time-independent variables f_* (the phase space distribution function), ρ_* (the smoothed density), and r_* (the radius); f_* is plotted against the homologous energy E_*, while ρ_* and r_* are plotted against the homologous potential, U_*. The dashed line shows f_* for the lowered Maxwellian distribution, $\exp(-E_*) - 1$.

varies as $\rho_h^{1/2} \propto \rho_G^{1/2}$, and will be relatively low for clusters which orbit the Galaxy at large mean distances, giving a low $\rho_G^{1/2}$.

Hénon's elegant solution of the equations for an evolving cluster can provide a realistic model for a globular cluster only if an energy source of the proper magnitude is present at the cluster center. As shown in chapter 7, interactions involving binary stars may produce such a source of energy.

Another important result obtained with the Hénon model is the dependence of escape rate on stellar mass. To evaluate this dependence, $\rho(r)$, $\phi(r)$, m_f and $f_f(E_f)$ were held fixed, taking the solution shown in Fig. 3.2. The Fokker-Planck equation (2-86) was then solved for stars with different

TABLE 3.1

Rate of Escape ξ_e for Stars of Different Masses

m/m_f	1.0	.8	.6	.4	.2	0
$100\xi_e$	4.5	6.6	8.7	11.0	13.4	15.8

values of m/m_f. Thus the solution represents the distribution of various stellar groups of different masses, each contributing a negligible fraction of the cluster mass, but each interacting with the predominant population of field stars, all with the same mass m_f. The final values for the escape probability ξ_e are given in Table 3.1. These values are referred to t_{rh} for stars of mass m_f, and do not refer to the longer interaction time between these lighter, more rapidly moving stars and the field stars. Stars heavier than $1.5m_f$ do not escape at all, since equipartition tends to make them all collect at the cluster center.

A much steeper increase of ξ_e with decreasing mass is found if a small population of light stars is assumed present in the uniform cluster considered in §3.1. For this uniform model ξ_e increases by a factor of 18 as m/m_f drops from 1 to 0, as compared to the factor 3.5 shown in Table 3.1. This difference is due at least in part to the fact that in a realistic cluster, with a strong concentration of stars in the central potential well, the less massive stars tend to be more uniformly distributed throughout the cluster than are the more massive objects, and spend relatively less time in the high-density regions where collisions are frequent. This effect does not occur in a hypothetical uniform cluster.

3.3 COLLAPSE OF AN ISOTHERMAL SPHERE

The instability of an isothermal region within the cluster and the process of collapse which it produces are major effects in cluster evolution. The basic cause of this instability is the negative specific heat of gravitationally bound systems. According to the virial theorem, equation (1-10), the kinetic energy T is half of the binding energy in an isolated system. Thus if the system loses energy, W becomes more negative and T increases; the system becomes hotter as it loses energy and cools as it is heated. If such a bound system is in thermal contact with a heat sink at some constant temperature, an instability can result, with heat flowing into the sink and the kinetic energies of the particles in the gravitationally bound system increasing steadily as the system contracts, losing energy.

Within a single cluster this process can be important if the central core is so concentrated compared with the rest of the system that its dynamical equilibrium is practically that of an isolated system. The core is in thermal contact with the outer regions of the cluster, which can serve as a heat sink. Thus it is possible for the core to lose energy to the outer regions, and to contract and heat up in the process. The increase of mean square random velocity in the core then encourages additional flow of heat from the core to the surrounding regions, increasing the rate of core collapse. This process, which involves the development of thermal gradients within the cluster and gravitational collapse within the core, is called the "gravothermal instability."

The analysis of such an instability is best carried out if the cluster is initially isothermal. Since, as we have seen in §1.2, an isolated isothermal cluster has an infinite mass and radius, a bounded isothermal sphere is usually considered, confined by a non-conducting rigid spherical shell of radius R. As one might expect, the possible appearance of the gravothermal instability depends on the density contrast $D \equiv \rho(0)/\rho(R)$, where $\rho(0)$ is the central density and $\rho(R)$ is the density just inside the confining shell. If D does not much exceed unity, the self-gravitational attraction of the core is a relatively small effect, and an increase of temperature within the central core will require an energy input rather than releasing energy. Only for sufficiently large D does the self-attraction of the core sufficiently outweigh the pressure of the surrounding layers so that for the mass in the core $T \approx -W/2$, and the core specific heat becomes negative; in this latter case core collapse will liberate heat, but increase $v_m^2(0)$, the central velocity dispersion, producing a runaway collapse.

We first discuss the criteria for the appearance of this gravothermal instability. The nature of the resulting collapse is then explored with a self-similar solution. While this process is known to occur when the mean free path is either much longer or much shorter than the spherical radius R, most analyses have assumed a short mean free path, considering the instability of an isothermal gas, with a distribution of velocities that is locally Maxwellian and hence isotropic.

a. Criteria for gravothermal instability

We consider now a spherical volume of radius R, occupied by a gas of mass M, with potential energy W, internal heat energy T, total energy $E_T = W + T$ and entropy S. The mass, volume and total energy are assumed constant. The system will be in mechanical equilibrium if any adiabatic displacement of fluid elements gives no change of energy $W + T$ to first order in the relative change of p, ρ and T. The equilibrium will be mechanically stable if to second order the energy $W + T$ increases for all such displacements; thus at a stable equilibrium point the value of $W + T$ must be a minimum against all such changes. Similarly, the system will be in thermal equilibrium if any flow of heat from one element to another produces no change in entropy to first order in $\Delta T/T$. The equilibrium will be stable if to second order the entropy decreases as a result of any assumed heat flow with E_T constant. Hence at a point of stable thermal equilibrium, S must be a maximum against all thermal changes which keep the total energy constant.

This situation is summarized in Fig. 3.3, which plots energy and entropy against one of the various parameters which characterize the state of the gas; in this case, we use the parameter D, the density contrast between the center and the outermost layer, at $r = R$. The horizontal scale plots log D, while the vertical scale represents the total cluster energy E_T in units of GM^2/R.

The solid line gives a plot of total energy E_T [4] for a one-parameter or linear series of equilibrium configurations, all of the same mass and vol-

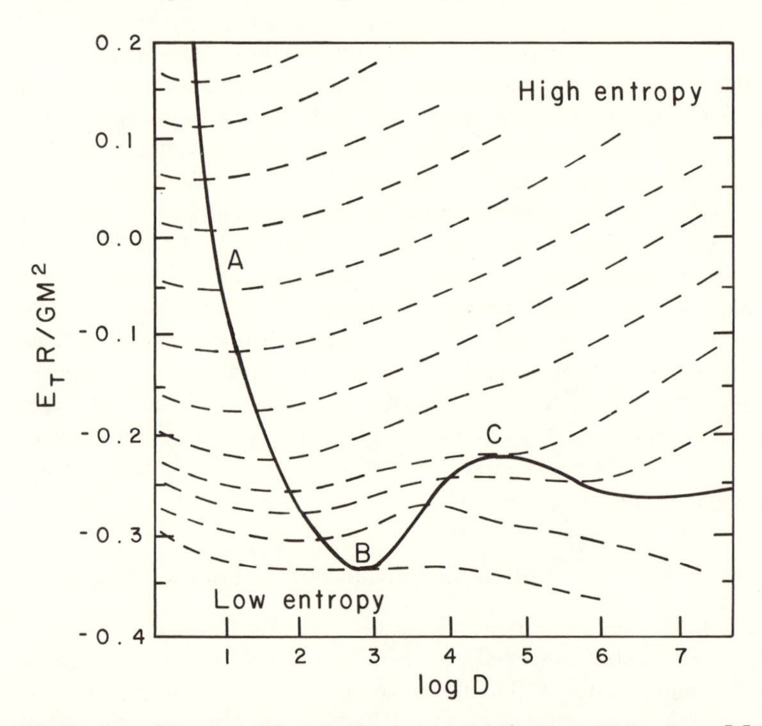

Fig. 3.3. Energy and Entropy of Bounded Isothermal Sphere. The solid line shows [4] the total energy, E_T, of an isothermal cluster of fixed mass, confined in an equilibrium state by a rigid shell of fixed radius R; E_T is plotted against $\log_{10} D$, where D is the density ratio between the center and the outermost region, at $r = R$. The entropy is constant along the dashed lines, which are qualitative only.

ume but of different E_T, T, S and D. Since the change of entropy dS along this sequence is given by dE_T/T, it is evident that E_T and S decrease together as D increases, and at point B, where $D = 708.6$ and E_T is a minimum, S is also a minimum. The dashed contour lines of constant S show qualitatively how S is altered if D is changed (with E_T still held fixed) with the system moving away from equilibrium.† For higher E_T, at point A, for example, these lines are concave upward, indicating that S decreases for any departure at constant E_T from the equilibrium state. On the other hand, as minimum energy is approached the dashed lines become more complex. The requirement that these curves be horizontal whenever they cross the equilibrium sequence line requires that their curvature reverse, as shown in the figure. It follows that the equilibrium sequences between B and C must be thermally unstable, since the entropy curves for fixed E_T become convex upward in this region. This method of linear-series analysis was originated by Poincaré and has since been developed by others [4,5].

In general the equilibrium at which some system variable reaches a maximum or a minimum value is called a "turning point." It will be noted that the appearance of a turning point in a linear series of equilibrium configurations indicates not which configurations are stable and which are unstable, but rather the boundary points between stability and instability. For a confined isothermal sphere, the sequences for low D, in the neighborhood of point A, correspond to a gas held in by the rigidity of the confining shell, with gravitational self-attraction negligible. Such configurations are clearly stable in the limit of low D, and hence the sequence between B and C must represent unstable configurations. There are certain circumstances where the indicated change of stability at turning points must be viewed with caution [6].

Indication of the gravothermal instability for a confined isothermal sphere with D exceeding about 709 was first obtained [7] by a direct computation of $\delta^2 S$, the second-order change of entropy, for perturbations at constant energy from an equilibrium state. For D exceeding the critical value, the detailed analysis shows that a positive change in entropy can occur, and instability must result. For smaller D the cluster is thermally stable.

When the mean free path of the confined particles much exceeds R, the analysis becomes more complicated. In particular, for most types of perturbations from equilibrium, a radial gradient of temperature appears and the perturbed $f(\mathbf{v},t)$ can no longer be isotropic. The computation of the

† The changes of entropy shown in Fig. 3.3 differ markedly from those in Fig. 2 of [4], which are drawn for fixed E_T and varying confining radius R.

entropy change in such situations is difficult. One conclusion that follows simply is that point B, where E_T is a minimum, remains a position of marginal stability for long mean free path as well as for short. The marginal perturbations in this case represent a drift along the equilibrium sequence, with constant v_m^2 and isotropic $f(\mathbf{v},t)$ and should be equally possible for systems with long or short mean free paths.

It should be emphasized that the unstable equilibrium configurations, beyond point B in Fig. 3.3, are not easily reached in a normal evolutionary process. If, for example, energy is withdrawn from the cluster by some process, so that its evolutionary point in Fig. 3.3 tends to fall steadily lower, the consequence is not that the cluster will reach an unstable state but rather that it will reach a situation where no equilibrium is possible for fixed E_T, M and V. Thus collapse will ensue, though with a time scale determined by the rate of heat flow, not by the dynamical time scale.

While it is customary to discuss the collapse produced by gravothermal instability, in principle an expansion is also possible. A cluster in unstable equilibrium between points B and C in Fig. 3.3 can move either to the right on this diagram, with D increasing and the core collapsing, or to the left, with the core expanding until the system reaches a stable equilibrium between points A and B. In actual clusters, core contraction is produced initially by evaporation and mass stratification, leading presumably to gravothermal collapse. As we shall see in §7.3, during the post-collapse phase of an evolved cluster, core expansion may at times be driven by gravothermal instability.

b. Model for gravothermal collapse

We consider now the detailed collapse process, following the onset of the gravothermal instability. As in §3.2, we look for a self-similar solution [8], and assume that all the physical variables are functions of the scaled radius $r_* = r/r_c(t)$, where r_c is the core radius. Hence we write

$$\begin{aligned} \rho(r,t) &= \rho_c(t)\rho_*(r_*), \\ v_m^2(r,t) &= v_c^2(t)v_*^2(r_*), \\ M(r,t) &= M_c(t)M_*(r_*), \end{aligned} \tag{3-28}$$

where $M(r,t)$ denotes the mass within the shell of radius r at the time t. We omit here the subscript m from v_c and v_*, whose product gives the rms random velocity. All numerical constants and dimensional factors in ρ, v_m^2 and M are absorbed in ρ_c, v_c^2 and M_c. While $\rho_c(t)$ and $v_c^2(t)$ are the densities and velocities at $r = 0$, $M_c(t)$ equals a numerical constant times the mass within the radius $r_c(t)$; to remove an arbitrary constant, $r_c(t)$ is set equal to κ, the reference radius for the core of an isothermal sphere—see equation (1-24). (Here $r_c(t)$ is about one-third of the core radius r_c defined in §1.1.)

We assume that the rate of core evolution is determined by the central relaxation time, $t_r(0,t)$, an assumption verified below. Introducing the collapse rate, ξ_c—analogous to the evaporation probability ξ_e in equation (3-3)—we write

$$\frac{1}{\rho_c}\frac{d\rho_c}{dt} = \frac{\xi_c}{t_{rc}}, \tag{3-29}$$

where t_{rc}—an abbreviated designation for $t_r(0,t)$—is again obtained from equation (2-61). In addition, the core kinetic energy E_c—analogous to the total kinetic energy ME_k in equation (3-7)—is assumed to be given by

$$E_c \propto M_c^{\zeta}. \tag{3-30}$$

As a result of assumptions (3-29) and (3-30), the self-similarity solution shows the same time dependences found in §3.1 for the evaporating uniform sphere. In particular, equation (3-30) is equivalent to equation (3-4), with M and E_T replaced by the core mass and kinetic energy, M_c and E_c; the gravitational energy, not included in E_c, varies in proportion to E_c during self-similar evolution. Equations (3-5) through (3-8) are still valid with all physical properties referring to the core, except that t_{coll} in equation (3-6) is now defined in terms of the constant ξ_c in equation (3-29).

This constant may be related to the quantity ξ_e introduced in §3.1, if in the defining equation (3-3) M and t_r are replaced by the core mass M_c and the central relaxation time t_{rc}, respectively. We then obtain from equations (3-29), (3-8) and (3-6)

$$\xi_c = (5 - 3\zeta)\xi_e. \tag{3-31}$$

Of course in the present context, ξ_e measures the rate at which the core mass decreases, not the rate at which stars escape from the cluster.

To determine the constant ζ in equation (3-30) requires a full solution of the equations of motion, as applied to the homology variables, ρ_*, v_*^2, M_* and r_*. Two additional assumptions are required for a solution.

The first assumption is that the disturbance is localized to the central regions of the cluster. This assumption is necessary if the collapse process is to be credible, since the outer regions are not clearly specified and almost certainly depart from the initial isothermal sphere whose instability was considered at the beginning of this section. Hence we assume that as r increases, $M(r,t)$ becomes independent of t. This condition yields

$$M_* M_c\left[\frac{1}{M_c}\frac{dM_c}{dt} - \frac{1}{r_c}\frac{dr_c}{dt}\frac{r_*}{M_*}\frac{dM_*}{dr_*}\right] = 0, \tag{3-32}$$

According to equation (3-5), $dr_c/r_c\,dt = (2 - \zeta)\,dM_c/M_c\,dt$. If for large r_*, we let

$$\rho_* \propto \frac{1}{r_*^{2+\beta}}, \tag{3-33}$$

equation (3-32) yields

$$\beta = (1 - \zeta)/(2 - \zeta). \tag{3-34}$$

Equations (3-33) and (3-34) provide a condition which the solution must satisfy at large r_*, if the collapse process is confined to the inner regions of the cluster.

The second assumption relates the heat flux F_h to dv_m^2/dr. To apply this assumption in a realistic way, we rewrite the coefficient of thermal conductivity in a form suitable for a mean free path much exceeding the cluster dimensions. For short mean free path the conductive heat flux F_h per unit area has the usual form

$$F_h = -b\rho v_m \lambda \frac{d}{dr}\left(\frac{3kT}{m}\right), \tag{3-35}$$

where λ is the mean free path and v_m^2, the mean square (three-dimensional) random velocity, equals $3kT/m$; the quantity b is a constant of order unity. The product $\lambda d(3kT)/m\,dr$ is the excess of v_m^2 between the starting point of the particles and their stopping point one collision later. For long mean free path this excess kinetic energy per unit mass may be computed with H replacing λ, where H is the scale height of the particle's orbit. However, the number of orbits required for one collision is t_r/t_d, where t_r is the relaxation time and t_d is the dynamical crossing time or period, about equal to H/v_m. Hence to obtain F_h in the situation of long mean free path, we must multiply the right-hand side of equation (3-35) by $(H/\lambda)(H/v_m t_r)$, giving

$$F_h = -\frac{b\rho H^2}{t_r}\frac{d}{dr}\left(\frac{3kT}{m}\right). \tag{3-36}$$

Finally we write for the relaxation time

$$t_r = \frac{v_m^3}{\alpha n \Gamma} = \frac{v_m^3}{\alpha 4\pi G^2 n m^2 \ln \Lambda}, \tag{3-37}$$

where $\alpha = 1.22$ in equation (2-61); Γ is given in equation (2-13). If in equation (3-36) we also replace $3kT/m$ by v_m^2 and equate H to the local Jeans length, which equals the scale distance, κ, for an isothermal sphere—see equation (1-24)—with local values used for ρ and v_m, we obtain the desired relationship,

$$F_h = -\frac{\alpha b \rho G m \ln \Lambda}{3 v_m}\frac{dv_m^2}{dr}. \tag{3-38}$$

In general, the divergence of the conductive heat flux F_h plus the work which the gas does by expansion under ambient pressure equals the decrease of the internal energy in the gas. If we write this energy conservation equation in terms of the homology variables $F_c(t)F_*(r_*)$, $v_c(t)v_*(r_*)$, etc. (omitting the subscripts h and m from F and v), we obtain an expression [8] in which various terms involve functions of t multiplied by functions of r_*. To obtain a solution we must assume either that the function of t in the various terms are all proportional to each other and cancel out, or that the various functions of r_* all cancel for a similar reason. We exclude this second possibility, since it does not satisfy the boundary conditions at $r_* = 0$ or ∞, and find

$$\frac{1}{r_*^2}\frac{d}{dr_*}(r_*^2F_*) = \rho_*v_*^2c_2\left[\frac{7-3\zeta}{2} - \frac{d\ln(v_*^3\rho_*)}{d\ln M_*}\right] \tag{3-39}$$

and

$$\frac{d\ln M_c}{dt} = -3c_2\frac{F_c}{\rho_c v_c^2 r_c}. \tag{3-40}$$

The constant first term in the brackets in equation (3-39) has been simplified with the use of equations (3-7) and (3-8). If we substitute into equation (3-40) the time dependent F_c obtained from equation (3-38) we obtain, making use of equation (3-37) and the relationship $r_c^2 = \kappa^2 = v_c^2/(12\pi G\rho_c)$ —see equation (1-24)—

$$\frac{1}{M_c}\frac{dM_c}{dt} = -\frac{3c_2b}{t_{rc}} = -\frac{\xi_e}{t_{rc}}. \tag{3-41}$$

where equation (3-3) for ξ_e has been modified as in the discussion leading to equation (3-31).

In addition to equation (3-39), the other three equations in the homology variables are

$$\frac{dM_*}{dr_*} = r_*^2\rho_*, \tag{3-42}$$

$$\frac{d(\rho_*v_*^2)}{dr_*} = -\frac{\rho_*M_*}{r_*^2}, \tag{3-43}$$

$$F_* = -\frac{\rho_*}{v_*}\frac{dv_*^2}{dr_*}. \tag{3-44}$$

Equations (3-42) through (3-44) are the equations of mass conservation, hydrostatic equilibrium, and thermal conductivity. Together with equation (3-39), these equations may be solved numerically subject to the four boundary conditions that $F_*(0) = M_*(0) = 0$, $\rho_*(0) = v_*(0) = 1$, together with the proper values for the two unknown constants c_2 and ζ. This latter variable determines the form of the solution, while the constant c_2, together with the value of the constant b in equation (3-38), determine the rate at which the cluster evolves per time interval t_{rc}, defined as the value of t_r at $r = 0$.

The solution obtained [8] gives $\zeta = 0.737$ and c_2 about equal to 0.96×10^{-3}. With this value for ζ, β equals 0.208 from equation (3-34), and equation (3-33) yields for large r_*

$$\rho_* \propto r_*^{-2.21}. \tag{3-45}$$

The detailed variation of ρ_* with r_* is displayed in Fig. 3.4. The curve falls somewhat below the values of $\rho(\xi)/\rho(0)$ for an isothermal sphere, with deviations less than 3 percent for $r_* < 14$, and less by some 30 percent as r_* increases to 100.

The temporal variations of M_c, v_c and ρ_c are related by equations (3-6) through (3-8) in §3.1. Thus for $\zeta = 0.737$ we obtain

$$\frac{v_c}{v_c(0)} = \left[\frac{\rho_c}{\rho_c(0)}\right]^{(1-\zeta)/(10-6\zeta)} = \left[\frac{\rho_c}{\rho_c(0)}\right]^{1/21.2}. \tag{3-46}$$

Also we find from equations (3-10) and (3-31)

$$\frac{t_{\text{coll}} - t}{t_{rc}} = \frac{2(5-3\zeta)}{(7-3\zeta)} \times \frac{1}{\xi_c} = \frac{1.165}{\xi_c}. \tag{3-47}$$

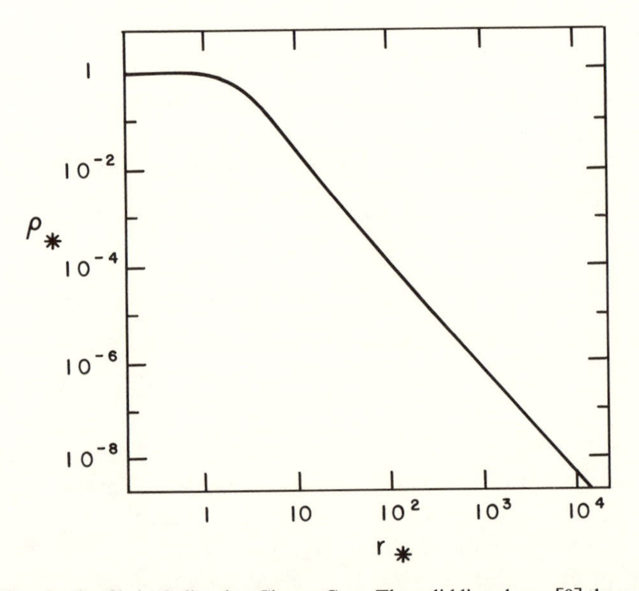

Fig. 3.4. Density Profile in Collapsing Cluster Core. The solid line shows [8] the variation of homologous density ρ_* (equal to $\rho(r,t)/\rho(0,t)$) as a function of r_* (equal to $r/r_c(t)$) for the core of a cluster undergoing collapse because of the gravothermal instability.

For the constant ξ_c which appears in this expression and also in equation (3-29), giving the temporal behaviour of ρ_c, we obtain from equations (3-31) and (3-41), together with the value cited above for c_2,

$$\xi_c = 8.0 \times 10^{-3}\, b. \tag{3-48}$$

The value of 3.6×10^{-3} found for ξ_c in §4.2c with detailed isotropic models indicates that the effective value of b in equation (3-38) equals 0.45.

Another result given by this self-similar solution is the fluid velocity $u(r,t)$. This quantity may be found from the condition that $M(r_*,t) = M_c(t)M_*(r_*)$ is constant with time if $dr/dt = u = u_*(r_*)\, dr_c/dt$. With straightforward computations, using equation (3-5), we find

$$u_* = r_*\left(1 - \frac{1}{2-\zeta}\frac{M_*}{r_*\, dM_*/dr_*}\right). \tag{3-49}$$

For low r_*, ρ_* is uniform and from equation (3-42) it follows that $M_*/(r_*\, dM_*/dr_*)$ equals 1/3; hence $u_*/r_* = 0.74$ and $u(r,t)$, like dr_c/dt, is inward. At $r_* = 16.1$, where ρ_* has fallen to about 0.0071, u_* vanishes and as r_* increases further, u_* first decreases to appreciable negative values and then increases to zero as equation (3-45) becomes asymptotically valid.

While this analysis does not demonstrate a clear transition from the onset of the gravothermal instability (or from a collapse initiated by the absence of any stationary equilibrium), the connection between these effects seems clear. The driving force for the collapse, as for the initial instability, is the negative specific heat of the inner regions, which heat up as they surrender energy during the collapse, providing an outward flux of heat.

3.4 MASS STRATIFICATION INSTABILITY

We consider next the equilibrium of a cluster in which the heavier stars become selectively concentrated in the inner regions as a result of the tendency toward equipartition of kinetic energy discussed in §2.1d. As the kinetic energy of the heavier stars approaches that of the lighter stars, the heavies will have lower random velocities and cannot rise so far out in the cluster against the gravitational potential $\phi(r)$. We examine here the conditions that must be satisfied for the heavies to reach an equilibrium situation in a cluster characterized by equipartition of kinetic energy.

To simplify the discussion, stars of only two masses, m_1 and m_2, will be considered; we assume $m_2 > m_1$. The total cluster mass in light and heavy stars will be denoted by M_1 and M_2, respectively. Two limiting cases are readily established. If M_2/M_1 is sufficiently small so that ρ_2 is everywhere negligible compared to ρ_1, then the potential ϕ is entirely determined by the lighter stars. In the inner regions, where conditions are nearly isothermal, ρ_1 will vary about as $\exp(-B\phi)$. If equipartition is established between the two mass groups, ρ_2 will vary about as $\exp(-m_2 B\phi/m_1)$, and the heavy stars will be more strongly concentrated to regions near the center, where ϕ is more strongly negative. By assumption, this central concentration of heavy stars toward the center does not increase ρ_2 up to ρ_1 and has no significant effect on ϕ and hence none on ρ_1 either.

The second limiting case is that in which M_2/M_1 is sufficiently large so that ρ_1 is everywhere negligible compared to ρ_2. The potential ϕ is now determined entirely by the heavier stars, whose mean square velocity will be one fourth the mean square v_e. If now equipartition is assumed and if m_2/m_1 exceeds 4, for example, the mean square velocity of the lighter stars will exceed the mean square escape velocity and most of these stars will promptly escape. Clearly, full equipartition is not feasible in this situation, and the tendency for equilization of kinetic energy will lead to a more rapid escape of stars significantly lighter than the average; according to Table 3.1, the increase in evaporation rate amounts to a factor of about 3 to 4 for very low masses.

The more interesting case, and the one of greatest importance for actual clusters, is the situation where $M_2 < M_1$, but where mass stratification increases ρ_2 in the central regions so that ρ_2 significantly exceeds ρ_1. As we shall see, full equipartition cannot be satisfied in this situation either, and the tendency towards equalization of kinetic energy leads to a steady collapse of the subsystem of heavy stars.

Instead of constructing detailed models, we shall use the Virial Theorem to obtain approximate results, making a variety of simplifying assumptions in the course of the analysis. If we apply equation (1-9) to the heavy stars in the cluster, we have

$$M_2 v_{m2}^2 = 4\pi \int_0^\infty \rho_2 r^3 \frac{d}{dr}(\phi_1 + \phi_2)\, dr, \tag{3-50}$$

where ϕ_1 and ϕ_2 are the gravitational potentials resulting from ρ_1 and ρ_2 respectively. We approximate this equation in two ways. The term in ϕ_2 leads to the total gravitational self-binding energy of the heavy stars, which we evaluate by means of equation (1-10). To evaluate the term in ϕ_1 we assume that ρ_1, the density of the lighter stars, has its central value $\rho_1(0)$ throughout the region occupied by the heavier stars. In the corresponding equation for $M_1 v_{m1}^2$ we neglect the contribution of ϕ_2, which is valid if M_2/M_1 is sufficiently small, and we again use equation (1-10).

If now we assume equipartition, $m_1 v_{m1}^2$ must equal $m_2 v_{m2}^2$ and we find, dividing through by 0.4,

$$m_2\left(\frac{M_2}{r_{h2}} + \alpha M_1 \frac{r_{h2}^2}{r_{h1}^3}\right) = \frac{m_1 M_1}{r_{h1}}, \tag{3-51}$$

where

$$\alpha \equiv \frac{5}{4} \frac{\rho_1(0)}{\rho_{h1}} \left(\frac{r_{m2}^2}{r_{h2}^2} \right). \tag{3-52}$$

The quantity r_{m2}^2 is the mean squared value of r for the heavy stars, which appears in evaluating the ϕ_1 term in equation (3-50). Following the notation used in previous sections, we have denoted by r_{h1} and r_{h2} the radii containing half the mass for stars of types 1 (light) and 2 (heavy), respectively. Also we define ρ_{h1} as the mean density for stars of type 1 within the radius r_{h1}, and similarly for stars of type 2. By definition we have

$$\left(\frac{r_{h1}}{r_{h2}} \right)^3 = \frac{M_1}{M_2} \frac{\rho_{h2}}{\rho_{h1}}, \tag{3-53}$$

and equation (3-51) assumes the form

$$\chi \equiv \frac{M_2}{M_1} \left(\frac{m_2}{m_1} \right)^{3/2} = \frac{(\rho_{h1}/\rho_{h2})^{1/2}}{(1 + \alpha\rho_{h1}/\rho_{h2})^{3/2}}. \tag{3-54}$$

The maximum value of χ as ρ_{h1}/ρ_{h2} varies is equal to $0.38\alpha^{-1/2}$. We denote this maximum value by $\chi_{\max}$; evidently M_2/M_1 cannot exceed $(m_1/m_2)^{3/2}\,\chi_{\max}$. In evaluating $\chi_{\max}$, we may set r_{m2}/r_{h2} equal to 1.13, the value for a Maxwellian distribution in the parabolic potential well resulting from a constant $\rho_1(0)$. In fact $\rho_1(r)$ decreases with increasing R, but if m_2 much exceeds m_1, the heavy stars will tend to be concentrated in the central region where ρ_1 is relatively constant. The ratio of $\rho_1(0)$ to ρ_{h1} is more widely variable, increasing from 2.5 to 4.4 for polytropes with n between 3 and 5. While this ratio is about one order of magnitude greater for the critical isothermal sphere, with $\rho(0)/\rho(R) = 709$, discussed in §3.3a, we shall here set $\rho_1(0)/\rho_{h1}$ equal to 3.5, giving $\alpha = 5.6$. Hence we obtain as a condition for equipartition in equilibrium,

$$\chi < \chi_{\max} = 0.16. \tag{3-55}$$

With the assumed value of α, $\chi = \chi_{\max}$ when $\rho_{h2} = 11.2\rho_{h1}$.

This result is valid when m_2 is appreciably greater than m_1, so that the heavy stars in equipartition are confined to the region of the core where $\rho_1(r)$ is nearly equal to $\rho_1(0)$. Also, in this limit M_2/M_1 is very small so that the effect of ϕ_2 on the distribution of most of the lighter stars is relatively small. As m_2/m_1 decreases toward unity, $\chi_{\max}$ will increase.

What happens when condition (3-55) is not fulfilled? In that case the heavy stars will continue to lose kinetic energy to the lighter ones and consequently fall in toward the center, thereby increasing their kinetic energy and increasing the departure from equipartition. According to equation (2-60), if $v_{mt} = v_{mf}$, the time constant $(\bar{E}_{Sf} - \bar{E}_{St})/(d\bar{E}_{St}/dt)$ for this process is about $m_f v_m^3/(m_t n_f \Gamma)$, about equal to $m_f t_r/m_t$, where t_r is for the field stars interacting with each other and is given in equation (2-61). Thus approach to equipartition is initially much more rapid than evaporation, which requires time intervals of about $100 t_{rh}$. However, this collapse of a heavy-star subsystem is somewhat self-limiting, since when the heavy stars far outnumber the light ones in the central core, the relative number of these lighter stars is inadequate to carry energy rapidly away from the heavier stars. These stages of runaway mass stratification are not subject to simple analyses, since no natural self-similar solutions have been found.

Model clusters have been constructed which violate condition (3-55) by substantial factors. However, an examination of the mean square stellar velocities in the gravitationally bound models of this type shows that most of the heavy stars are far from equipartition with the surrounding lighter stars. The method of turning points in a linear series of equilibria, discussed in §3.3a, has been applied [9] to the stability of two-component models, based on a velocity distribution function similar to the lowered Maxwellian. The results show that if equipartition is established with sufficient precision (mean kinetic energies of the two masses equal to within some 15 percent at the cluster center), the critical value of χ for stability is between 0.16 and 0.20. Stability for appreciably higher values of χ requires significantly higher deviations from equipartition even at the center. Thus such an equilibrium, if made possible with appreciable deviations from equipartition, tends to be unstable.

REFERENCES

1. L. Spitzer and R. Härm, *Ap. J.*, **127**, 544, 1958.
2. I. R. King, *A. J.*, **70**, 376, 1965.
3. M. Hénon, *Annales d'Astroph.*, **24**, 369, 1961.
4. D. Lynden-Bell and R. Wood, *M. N. Roy. Astr. Soc.*, **138**, 495, 1968.
5. J. H. Jeans, *Astronomy and Cosmogony* (Cambridge Univ. Press), 2nd edition, 1929, chap. VII.
6. G. Horwitz and J. Katz, *Ap. J.*, **222**, 941, 1978; and J. Katz, *M. N. Roy. Astr. Soc.*, **183**, 765, 1978.
7. V. A. Antonov, *Vestnik Leningrad Univ.* (Series Math. Mech. and Astron.), **7**, 135, 1962. English translation in *Dynamics of Star Clusters*, IAU Symp. No. 113, 1985, ed. J. Goodman and P. Hut (Reidel, Dordrecht), 1985, p. 525.
8. D. Lynden-Bell and P. P. Eggleton, *M. N. Roy. Astr. Soc.*, **191**, 483, 1980.
9. J. Katz and L. G. Taff, *Ap. J.*, **264**, 476, 1983.

4

Dynamical Evolution of the Standard Model

We discuss now the detailed dynamical evolution which the velocity perturbations of chapter 2 produce in the zero-order steady solutions of §1.2. The three basic approximations of §1.2 are, of course, retained for the zero-order solutions; i.e., (A) a smooth gravitational potential, averaged over the granularity of the stellar distribution, (B) a quasi-steady state for the cluster, with $f(\mathbf{r},\mathbf{v})$ and other physical properties changing slowly as the cluster evolves through a succession of essentially steady states, and (C) complete spherical symmetry in $\mathbf{r}$, with $f(r,v_r,v_t)$ independent of the direction of the transverse velocity $\mathbf{v}_t$.

The first section below discusses the basic model building methods that have mainly been used in following cluster evolution produced by small perturbations. In fact, these methods have mostly been applied to the perturbations produced by distant encounters between pairs of passing stars, which produce diffusion of stars in velocity space; we refer here to the resulting models as "standard." Among the perturbing effects that we exclude from the standard models are those associated with binary stars, finite stellar radii, and fluctuating gravitational fields due to neighboring masses outside the cluster. On the other hand, we include in the standard cluster models the presence of stars with a variety of masses. This arbitrary definition has the advantage that it includes most of the detailed model computations carried out before 1984. This chapter and the subsequent one together cover most of the important physical effects that occur in a cluster before core collapse reaches its climax.

While the discussion of detailed methods in §4.1 is focussed on evolution of the standard model, resulting from encounters between pairs of passing stars, the treatment is relevant also to other types of perturbations, whose effect on the zero-order solutions is analyzed in chapters 5 and 7. However, the discussion of results in §4.2 is limited to phenomena which are characteristic of the standard model and which, for the most part, have already been treated with idealized models in chapter 3.

4.1 METHODS

Several different techniques have been developed for computing cluster evolution caused by diffusion in velocity space. Before we discuss these, we mention briefly two entirely different methods which lie somewhat outside the framework presented here.

The first of these is the use of fluid dynamics, with the cluster treated as a perfect gas. This approach has the great advantage of simplicity. The fluid equations are well known and need no review here. These equations have been used in various analyses of clusters; in particular, they have given very useful results for the collapse and subsequent expansion of the cluster—see §§3.3b and 7.3. Since an actual cluster is characterized by very long mean free paths, much exceeding the cluster dimensions, it is not clear *a priori* what confidence can be placed in models based on a short mean free path. Hence any results obtained from the fluid equations must be regarded as somewhat provisional until confirmed with more accurate models.

Another alternative method, which dispenses with most of the simplifying assumptions made in §§1.2 and 2.2, is direct integration of the dynamical equations for the N mass points in the cluster. This method, sometimes referred to as "N-body calculations," can readily be applied to almost any situation [1], including transient perturbations by other masses located with no simple symmetry. The great disadvantage of this technique is the prohibitive amount of computing time required if N is large. In particular, computing the potential gradient at each star and following stars that approach close to their neighbors requires many operations.

To achieve greater efficiency in such calculations a variety of special techniques have been developed [2]. The singularity that develops in the equations as R_{ij}, the distance between stars i and j, approaches zero may be eliminated by a change of variables from $\mathbf{r}_i$, $\mathbf{r}_j$ and t to other suitably chosen functions, a process known as "regularization." The time steps used may be different for different stars, depending on the relative rates at which the distances to their closest neighbors are changing. Finally, the potentials between stars with large separation can be evaluated less frequently than for closer stars. With these methods, clusters with N exceeding 1000 have been followed numerically, though in most such N-body calculations N has been between 100 and 500.

While calculations for systems with 100 to 1000 stars provide an interesting comparison with models based on a large-N theory, the two sets of calculations are not strictly comparable, since the structure and evolution of a cluster vary with N in three different ways. With decreasing N the relative importance of close encounters becomes greater, as may be shown directly from the results in §2.1a; hence the validity of the Fokker-Planck equation becomes doubtful. In addition, as N decreases, the ratio of the dynamical time, $t_{dh} \equiv r_h/v_m$, to t_{rh} becomes larger; with the same assumptions as led to equation (2-63) we find.

$$\frac{t_{dh}}{t_{rh}} = \frac{11 \ln (0.4N)}{N} \tag{4-1}$$

An increase in this ratio increases the energy of the escaping stars—see §4.2b—and thus affects the evolutionary track of the cluster. Finally, the rate of binary formation by three-body encounters in the cluster core, per relaxation time, varies as $1/N_c$, where N_c is the number of stars in the core—see §7.1; hence binaries form at progressively earlier evolutionary times as N decreases. For these reasons N-body calculations for an entire cluster are not discussed here.

As computers of greatly increased capability become available, N-body calculations may become applicable to systems with N as great as 10^5. Such calculations should provide a very useful tool for analyzing the more complex evolutionary stages of a cluster, especially those which depart from spherical symmetry. In the meantime, the primary application of such techniques is to the dynamics of a collapsing core when N_c, the number of stars within the core radius r_c, falls below 100. In this situation, which can arise at the climax of core collapse, the Fokker-Planck equation becomes invalid and also the formation of binary stars by three-body encounters becomes important. Section 7.2a describes a hybrid code developed to analyze this phase of cluster evolution; N-body calculations are made for stars within or near the core, while for the bulk of the stars at larger distances diffusion in velocity space is assumed.

Aside from this special but important situation, and perhaps also from more complex situations which may develop after core collapse, the velocity diffusion produced by distant two-body encounters plays the dominant role in the evolution of globular clusters. We turn now to a discussion of three principal methods used to analyze the effects of this diffusion on the standard model cluster, defined above.

These methods include two different Monte Carlo techniques and a numerical solution of the Fokker-Planck equation. While these approaches are, in principle, nearly identical, since they are all based on the physical processes analyzed in §§1.2 and 2.1, the techniques they use to compute the changes in the velocity distribution function are quite different; each method has its own particular advantages and disadvantages.

a. Dynamical Monte Carlo method

The Monte Carlo technique provides a straightforward method of computing changes in the velocity distribution function as a result of encounters. In this approach, a large number of particles is represented by a relatively small subset, each of whose members is followed through time with a computer; each of these particles will experience encounters at known rates, producing effects which are chosen at random, in accordance with known probabilities. Thus in the case of stars in a cluster, a small fraction of stars, which we shall call test stars, can be followed in time and their velocities perturbed at regular intervals by amounts $\Delta\mathbf{v}$, where the probability of each $\Delta\mathbf{v}$ is computed from the theory of two-body encounters. If N_t, the number of stars followed in this way, is as great as 10^3, about the value generally taken, the evolution found for the distribution function of test star velocities may be a fair representation for all stars in the cluster.

If each value of $\Delta\mathbf{v}$ considered for a particular test star corresponded to a single encounter with some passing star, the number of encounters to be considered would be extremely large. For example, $\Delta v_\perp$ is appreciable in each encounter, but $\langle \Delta v_\perp \rangle = 0$ because of symmetry; to obtain this result approximately by summing over various encounters would evidently require many such encounters for each stellar position. Hence, the values of $\Delta\mathbf{v}$ actually applied in a Monte Carlo program are each taken to represent an average over many successive encounters. The various Monte Carlo methods developed for studies of cluster evolution use different types of averages.

In some ways the most straightforward application of the Monte Carlo technique, and the one we consider first, involves a dynamical integration of the equations of motion for each of the N_t test stars, with $\Delta\mathbf{v}$ and $(\Delta v)^2$ chosen to represent proper averages over all types of encounters at each orbital position. Hence $\Delta\mathbf{v}$ is obtained directly from the diffusion coefficients obtained in §2.1. For example, if a star is moving in a radial orbit, with a velocity v_r along the cluster radius, but with zero transverse velocity v_t, and if a velocity perturbation is applied after each time interval Δt_p, we may compute the changes Δv_r and Δv_t from the equations

$$\begin{aligned} \Delta v_r &= \langle \Delta v_{||} \rangle \Delta t_p + X[\langle (\Delta v_{||})^2 \rangle \Delta t_p]^{1/2}, \\ \Delta v_t &= Y[\langle (\Delta v_\perp)^2 \rangle \Delta t_p]^{1/2}, \end{aligned} \tag{4-2}$$

where X and Y are random numbers with mean values of zero and mean square values of unity. The three diffusion coefficients are functions of v and of the velocity distribution of the field stars. For stars with $v_t = 0$ initially equations (4-2) give values of Δv_r and Δv_t with correct first and second moments. More general equations are readily obtained [3] for test stars moving in arbitrary directions. If we assume that the velocity distribution of the field stars is isotropic and Maxwellian, the diffusion coefficients may be taken from equations (2-52) through (2-54), with j^2 obtained from equation (2-11) and from the mean square velocity of the test stars in the adjacent regions of the cluster (i.e., with adjacent values of r, the distance from the center).

We consider next the dynamical integration for each test star, in the interval between velocity perturbations. The chief problem here is to determine the gravitational potential, $\phi(r)$. For this purpose the radial distribution of test stars is taken to be representative of that for all the cluster stars. One may think of each test star as being representative of 100 stars, for example, all with the same values of r, v_r and v_t, and distributed uniformly over a spherical surface. Thus for $N_t = 10^3$ the cluster is composed of 1000 "superstars," each a shell composed of 100 stars, if $N = 10^5$ for the entire cluster. We give each shell an integral number j, which we call the "rank," in order of increasing r_j, with $j = 1$ for the smallest shell. If we denote by r_j the radius of the superstar of rank j, the basic dynamical equation is

$$\frac{d^2 r_j}{dt^2} = \frac{J_j^2}{r_j^3} - \left(j - \frac{1}{2}\right)\frac{GM_s}{r_j^2}, \tag{4-3}$$

where J_j is the angular momentum per unit mass, equal to $r_j v_{tj}$, for the shell of rank j, and the factor 1/2 takes into account the diminished attraction of shell j on itself. The quantity M_s is the mass of each superstar, equal to M/N_t. This combination of dynamical integrations with Monte Carlo velocity perturbations [3] is sometimes referred to as the "Princeton method."

The procedures described above do not conserve energy very accurately. Changes in total energy arise because statistical fluctuations appear with the velocity perturbations, and also because changes in field-star energy, discussed in §3.1b, are ignored. To preserve energy conservation, the velocity perturbations applied to each group of 40 adjacent shells are adjusted to give zero change of kinetic energy for each group as a whole. Energy changes can also result from numerical integration of equation (4-3) when shells cross each other, when shells pass through pericenter or apocenter, or when shells approach very close to the origin. In such cases, the usual energy integral is used after each integration time step to adjust v_j^2. To avoid the singularity at small r_j in equation (4-3), a reflecting sphere of small radius (about one percent of r_h, the half-mass radius) is generally assumed in the Princeton models. With these procedures, the total energy change from start to finish for each model is generally less than one percent.

The Princeton Monte Carlo method has two advantages. Use of equation (4-3) makes it possible to examine violent relaxation, which may be associated with the initial stages of cluster evolution. More important, the correct treatment of the velocity changes produced in a single orbit makes it possible to investigate the rate of escape from an isolated system. The chief disadvantage of this technique is that it requires more computing time than other methods, since the dynamical integrations, while relatively straightforward, require from 1000 to 3000 integration time steps for each interval of t_{rh}.

b. Orbit-averaged Monte Carlo method

In this general approach one computes for each test star the change with time of the two orbital parameters, the energy E and angular momentum J, each taken per unit mass. Thus the time consuming dynamical integrations around each orbit are not required. As in the dynamical method, the number N_t of test stars considered is about 1000. For computation of the potential, each test star may again be regarded as a superstar, composed of many stars of identical r, v_r and v_t.

Two methods along this line have been developed and extensively applied. The first of these, the earliest Monte Carlo method applied to clusters, is generally known as the "Hénon method" [4,5] and will now be described. In this method at time intervals Δt, taken to be less than $0.01 t_{rh}$, the test stars are assumed to have gravitational encounters in pairs, leading to changes in the velocity of each, with equal and opposite changes in energy and also in angular momentum of the two stars involved.

The first step in each cycle in the cluster's evolution is to determine the position r and the velocities v_r and v_t for each test star. The position is found from the potential obtained in the previous cycle. Specifically, values of r_p and r_a, the minimum (pericenter) and maximum (apocenter) distances from the center, are determined [4] from $\phi(r)$, E and J. The actual radial distance of the test star (and of its accompanying shell or superstar) is determined at random between r_p and r_a, weighting each dr by $1/v_r$, in proportion to the time spent in that radial interval. As a result of this procedure, when many successive encounters are included $\langle \Delta E \rangle$ is correctly averaged over the test-star orbit, as in equation (2-91). Once a value of r is selected, $v_t = J/r$, and v_r^2 is found from $\phi(r)$, v_t and E. As in the dynamical method, the superstars are then ranked by j, in order of increasing r. Once the rank of every superstar is known, the potential function $\phi(r)$ is redetermined. The resultant change of $\phi(r)$ at $r = r_j$ leads to an equal change in the energy E_j of each test star.

Next the velocity perturbations produced by encounters are computed. The superstars with $j = 1$ and 2 are assumed to have a gravitational encounter with each other, likewise those for $j = 3$ and 4, and so on up to $j = N_t - 1$ and N_t. Instead of accumulating the velocity changes in successive encounters with a variety of impact parameters, theory is used to integrate over the impact parameters of all encounters during the time step Δt with the same relative velocity V, and thus to determine the mean square cumulative value of the deflection angle χ in the relative orbit during Δt. The effective impact parameter p is then taken to give this cumulative χ^2 in a single Monte Carlo encounter. The number n_f of field stars per unit volume needed in this computation of χ is obtained from the radial separation of six adjacent superstars, centered at the interacting pair.

The analysis presented in §2.1b indicates that this use of a single p value to represent many encounters, each with a different value of p/p_0, gives correctly all the dominant diffusion coefficients in an inertial frame. These coefficients are all based on linear combinations of $\Delta v_{\|}$ and $(\Delta v_{\perp})^2$ in the center-of-mass frame for each encounter, and according to equations (2-18) and (2-19) the ratio of these two quantities is independent of p/p_0 for large p/p_0. Hence, consideration of encounters all with the same suitably chosen values of p/p_0 and χ gives correct dynamical results.

In the determination of the cumulative χ^2 and the effective p/p_0, the equation for $\chi/2$, obtained from equations (2-4) and (2-5), is approximated by

$$\frac{p_0}{p} = \tan\frac{\chi}{2} \approx \frac{\chi}{2}. \tag{4-4}$$

While this approximation is valid for actual distant encounters, it is somewhat inaccurate for the effective value of χ used in a Monte Carlo encounter, and a small correction [5] was made for this effect.

Once an effective impact parameter has been determined for the encounter, new velocities for the two superstars are computed, first in the reference frame of their center of gravity and then in the reference frame of the cluster. This transformation of reference frames involves a random choice of the angle between the orbital plane and a fixed plane; this angle is measured in a plane perpendicular to the relative velocity $\mathbf{V}$. The angle between the initial $\mathbf{v}_t$ vectors of the two superstars was selected at random earlier in the determination of V. The final superstar velocities, after each set of encounters between all pairs, then give new values of E and J for each superstar. The cycle is then repeated, starting with a redetermination of the superstar positions.

This procedure has been described with the simplification that the cycle time Δt is the same for all stars. The efficiency of this method is in fact increased [5] if Δt is varied with the position of the test star, so that Δt remains a small fraction of the local relaxation time. To achieve this objective, pairs of adjacent test stars (with $\Delta j = 1$) were selected at random for a gravitational encounter; the selection probability was assumed proportional to $1/t_r$ at the local density. After each new set of values for E and J and then for r and v had been found for each of the two interacting test stars, the j values for all stars, indicating their ranking, were redetermined as necessary.

This orbit-averaged Monte Carlo method has the great advantage that it requires much less computing than the corresponding dynamical method. Typically the number of time steps needed for evolution of one cluster averages only a few hundred per superstar. A scientific advantage of this approach is that the velocity perturbations are computed with the velocity distribution for the test stars; the test stars encounter each other, and the assumption of a Maxwellian velocity distribution for the field stars, which was incorporated in the Princeton method, is irrelevant here.

A modified version of this orbit-averaged Monte Carlo method has been developed by a group at Cornell [6,7] to give information on processes which occur on an orbital time scale, such as escape of stars or their capture by a central black hole. This approach, to which we refer as the "Cornell method," computes values of ΔE and ΔJ resulting from encounters during an integral number, n, of orbits, with n set equal to 1, for example, for stars nearing the escape energy. The Hénon method does not give such results directly, since an average is taken over encounters spread over many orbits to give statistically accurate results. Instead, the Cornell method uses the five orbital-averaged diffusion coefficients in equation (2-90), defined as in equation (2-91); these give directly the proper average rate of increase of E and J resulting from stellar encounters during one orbit, together with the average cumulative values per orbit of $(\Delta E)^2$, $(\Delta J)^2$ and $\Delta E\,\Delta J$.

Specifically, the equations used for ΔE and ΔJ are

$$\begin{aligned}\Delta E &= \langle \Delta E \rangle_{\text{orb}} \Delta t + X[\langle (\Delta E)^2 \rangle_{\text{orb}} \Delta t]^{1/2} \\ \Delta J &= \langle \Delta J \rangle_{\text{orb}} \Delta t + Y[\langle (\Delta J)^2 \rangle_{\text{orb}} \Delta t]^{1/2} \\ \Delta t &= nP(E,J),\end{aligned} \tag{4-5}$$

where each "orbital average" is defined as in equation (2-91), and $P(E,J)$ is again the orbital period. As in equations (4-2), X and Y are random numbers whose average vanishes, but whose mean square value is unity. In addition, X and Y are chosen to be correlated, giving

$$\langle XY \rangle = \langle \Delta E \Delta J \rangle_{\text{orb}} / [\langle (\Delta E)^2 \rangle_{\text{orb}} \langle (\Delta J)^2{}_{\text{orb}}]^{1/2}. \tag{4-6}$$

The diffusion coefficients involving E have been evaluated in §2.1c for an isotropic distribution of field star velocities. The remaining coefficients, involving J also, may be similarly expressed [6] in terms of $\langle \Delta v_{\|} \rangle$, $\langle (\Delta v_{\|})^2 \rangle$ and $\langle (\Delta v_{\perp})^2 \rangle$. In computing these diffusion coefficients the velocity distribution of the field stars is set equal to that of the test stars, suitably isotropized.

The time step $\Delta t = nP$, which differs for different test stars, is subject to the criterion that n must be an integer, but is otherwise as large as consistent with

$$[\langle(\Delta E)^2\rangle_{\text{orb}}\Delta t]^{1/2} \leqslant 0.15\,|E|,$$

and (4-7)

$$[\langle(\Delta J)^2\rangle_{\text{orb}}\Delta t]^{1/2} \leqslant 0.1J \text{ and also } 0.5(J_c - J).$$

Evidently the perturbations are computed orbit by orbit whenever the rms ΔE during one orbit is comparable with E, and similarly for ΔJ.

At intervals, $\rho(r)$ and $\phi(r)$ are recomputed, together with all the necessary orbit-averaged diffusion coefficients. In this computation of $\phi(r)$, as in all other Monte Carlo methods, the radial density distribution of the test stars is assumed identical with that of all the cluster stars. The change of E which results from the change of $\phi(r,t)$ and which must be added to ΔE in equation (4-5) is given by equation (2-93). In the other two Monte Carlo methods, equation (2-93) is satisfied automatically by the procedures used. Since in the Cornell method the energy distributions of the field and test stars are identical, this technique conserves energy except for statistical fluctuations, whose effect is removed by enforcing conservation of total cluster energy after each time step.

Another basic addition to the Monte Carlo program made by the Cornell group is a cloning procedure, designed to increase the number of particles in the central regions, where collapse produces a core with progressively fewer stars. For this purpose, whenever a test star decreases its energy below a certain boundary value, nine additional clones are created, each with the same properties initially as the original test star. Any clones recrossing this boundary are destroyed. Renormalization of the ratio between test stars and total stars is, of course, carried out [7] for energies below the boundary. Several such boundaries can be included. As a result of cloning, the number of test stars in a typical computation increased from its starting value of 500 to about 2000.

c. Numerical solution of the Fokker-Planck equation

The numerical solutions of the time-dependent Fokker-Planck equation, with N a function of E, J and t, have set $\partial N/\partial t$ equal to $(\partial N/\partial t)_{\text{enc}}$, as given in equation (2-90). Since the $\partial f/\partial E$ term in the central group of equation (2-81) is ignored, the change of E with time as ϕ slowly varies is taken into account through the constancy of the adiabatic invariant, $Q(E,J)$, defined in equation (2-92).

The Fokker-Planck approach gives more precise results than the Monte Carlo methods, but is based on the same assumptions; i.e., the three basic assumptions in §1.2 concerning the zero-order solution, and the assumption of small velocity perturbations, resulting from two-body encounters, discussed in chapter 2. As in the Cornell method, the diffusion coefficients are computed with a field star velocity distribution obtained from that of the test stars, isotropized with a suitable average of $f(E,J)$ taken over J. As pointed out earlier, one would expect this assumption to be realistic in the inner regions of the cluster, where most perturbations of the velocity occur and where $f(E,J)$ is found to be nearly independent of J. Additional evidence on the validity of this assumption is presented in §4.2c.

In practice [8], the variable $R \equiv J^2/J_c^2$ was used in the computations. As we have seen in §1.2, $J_c(E)$ is the angular momentum per unit mass of a star in a circular orbit, and is the maximum possible value of J for each E. Hence R varies from 0 to 1, independently of E, which has computational advantages. Equation (2-90) is readily modified with the substitution of $N(E,R)dE\,dR$ for $N(E,J)dE\,dJ$, and with the diffusion coefficients evaluated in E,R space. In the numerical computations $f(E,R)$ was advanced one step by means of a set of finite difference equations. Following this perturbation of $f(E,R)$, the potential $\phi(r)$ was recomputed by inverting Poisson's equation (1-5), with $f(Q,J)$ held invariant in the process. These steps kept the total energy E_T nearly constant, with a small drift amounting to $2 \times 10^{-4}\, E_T$ in a time interval equal to $t_r(0)$, the relaxation time at the center.

Greater accuracy was obtained in a separate program [9] in which f was assumed isotropic, independent of J. The rate of change of f due to encounters, $(\partial f/\partial t)_{\text{enc}}$, was taken from equation (2-86), with $\partial q/\partial t$ ignored. The change of E with time as a result of changing $\phi(r)$ was taken into account as in the anisotropic case, with the appropriate adiabatic invariant now given by q, defined in equation (2-83); $f(q)$ was held constant as $\phi(r)$ was recomputed. The secular energy drift rate was reduced by two orders of magnitude below its value in the $f(E,J)$ computation. A logarithmic radial mesh was used, and density increases were followed up to a factor 10^{20}, much exceeding the relevant values for a cluster with only 10^5 stars. By contrast, the density increase in the anisotropic model [8] was limited to a factor 10^3.

Since the time step used is unrelated to the period, this approach gives no information on the rate of escape from the isolated systems to which it has been applied. This detailed program for solving the time-dependent, Fokker-Planck equation has been generalized to include several mass components [10] and also certain interactions involving binary stars [11].

4.2 RESULTS

The techniques described in the previous section have been applied in numerous investigations of what is here called the standard cluster model—an isolated system with no perturbing external fields, no binaries and no physical changes in the stars. The results are summarized in this section and compared with the simplified theoretical models of the previous chapter. This discussion indicates clearly the relative importance during cluster evolution of the separate physical processes described in chapter 3.

a. Core-halo structure

All realistic calculations for the standard model show that the clusters, regardless of their initial conditions, soon develop two distinct regions, an inner isothermal sphere and an outer halo. In the inner sphere, which may contain typically half the cluster mass, the velocity distribution is nearly isotropic; while v_m decreases somewhat with increasing r, the density distribution $\rho(r)$ in this inner region is approximately that for an isothermal sphere. At the center of the cluster is a core of relatively uniform density with a radius r_c, as discussed in §1.1. Outside the isothermal region the rms transverse velocity, v_{tm}, decreases much more rapidly with increasing r than does v_{rm}, the corresponding rms radial velocity, and the stars in this surrounding halo move in predominantly radial orbits. This configuration is often called a "core-halo structure," although a massive and extended isothermal zone separates the core proper from the halo of stars moving nearly radially, which we shall call a "radial halo."

While encounters between pairs of stars are required to establish the isothermal region, if this does not exist in the initial state, a radial halo can be established during an early collapse phase if initially the system does not satisfy the Virial Theorem, equation (1-10); i.e., if $2T$ is substantially less than $|W|$. Calculations [12] with the dynamical Monte Carlo method, with no collisional perturbations of velocity considered, show that a cluster of initially uniform density, with $T = -W/4$ initially, develops a density profile with $d \ln \rho / d \ln r$ about equal to -3.8, much steeper than the value -2 for the outer regions of an isothermal sphere. Other calculations of initial collapse, taking into account some deviations from spherical symmetry as well as smaller-scale density fluctuations, show a variety of results [13], but in the halo the resulting $d \ln \rho / d \ln r$ is between -3.2 and -3.8 for about three-fourths of the models computed.

This acceleration of some stars to higher energies results from the crossing of shells occurring in the calculation. Those shells which initially move outwards will fall toward the center after most other stars, and will be subject to inwards gravitational acceleration by most of the cluster. However, they will reach their pericenter relatively late, after most shells have passed them, moving outwards, and during the "first bounce" of the system they will move outwards against the much weaker gravitational field produced by the few shells of smaller radius. Thus the total energy of some shells can be significantly increased during the collapse phase.

The density distributions obtained for two typical Monte Carlo cluster models [3] are shown in Fig. 4.1 at three evolutionary times, as measured by t/t_{rh}, where t is the elapsed time since the start of cluster evolution and t_{rh} is the initial half-mass relaxation time—see equation (2-63). Model D, shown on the left, starts out of virial equilibrium, and during the collapse phase r_h decreases from 79 to 43 dimensionless units, producing a marked radial halo at the very beginning. The collapse becomes singular at t/t_{rh} about equal to 18. Model F, on the other hand, starts life as a uniform sphere in equilibrium, with all stars moving in circular orbits and an outer radius of 66 units. Here the halo is absent initially and gradually develops as encounters tend to develop a velocity distribution which in the inner regions is more nearly isotropic and Maxwellian, with r_h for the entire system equal to 49. In this model the collapse is singular at $t/t_{rh} \approx 12$. At the latest time shown for each model, the time τ remaining before complete collapse is about $0.7t_{rh}$.

The end results are about the same for both models. In particular, for $r \leqslant r_h$, the density distribution is not far from the theoretical function for an isothermal sphere with the same central density, $\rho(0)$, and central rms

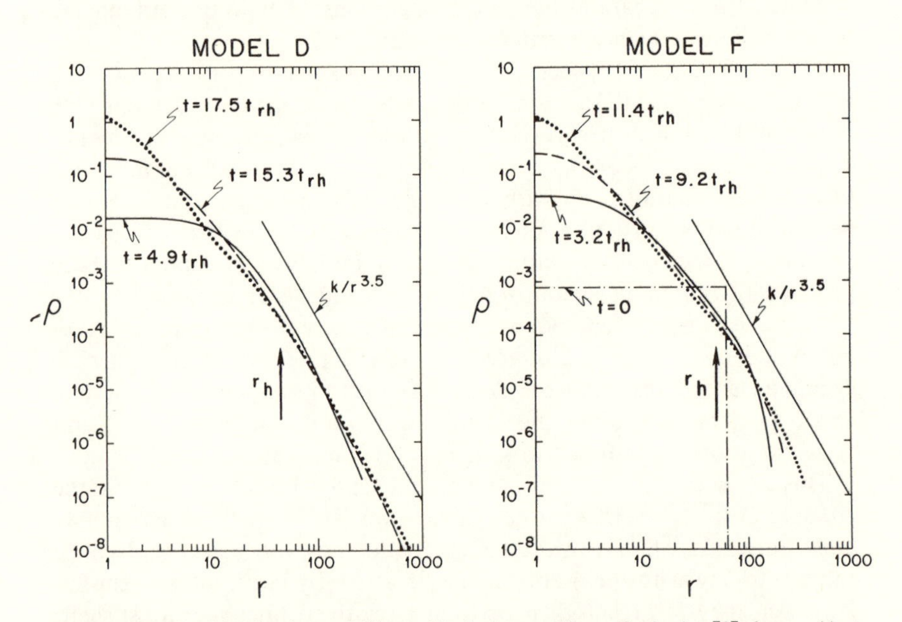

Fig. 4.1. Density Distribution at Different Evolutionary Times. Each plot [3] shows $\rho(r)$ at three different times (expressed here in units of t_{rh}) after the birth of the cluster. Model D began with an initial dynamical collapse, while Model F started with all stars in circular orbits and with $\rho(r)$ initially uniform, as shown by the dot-dash line. A variation of ρ as $r^{-3.5}$ is indicated, for comparison with the asymptotic form of the halo. The values of the median radius, r_h, during most of the life of each system, are also shown.

random velocity dispersion $v_m(0)$. The mean square radial and tangential velocities for these two models, at the same times as those shown in Fig. 4.1, are plotted against radius in Fig. 4.2. For $r \leqslant r_h$ the dots, representing the mean square tangential velocities v_{tm}^2, are about a factor of two higher than the mean square radial velocities v_{rm}^2, as would be expected for an isotropic

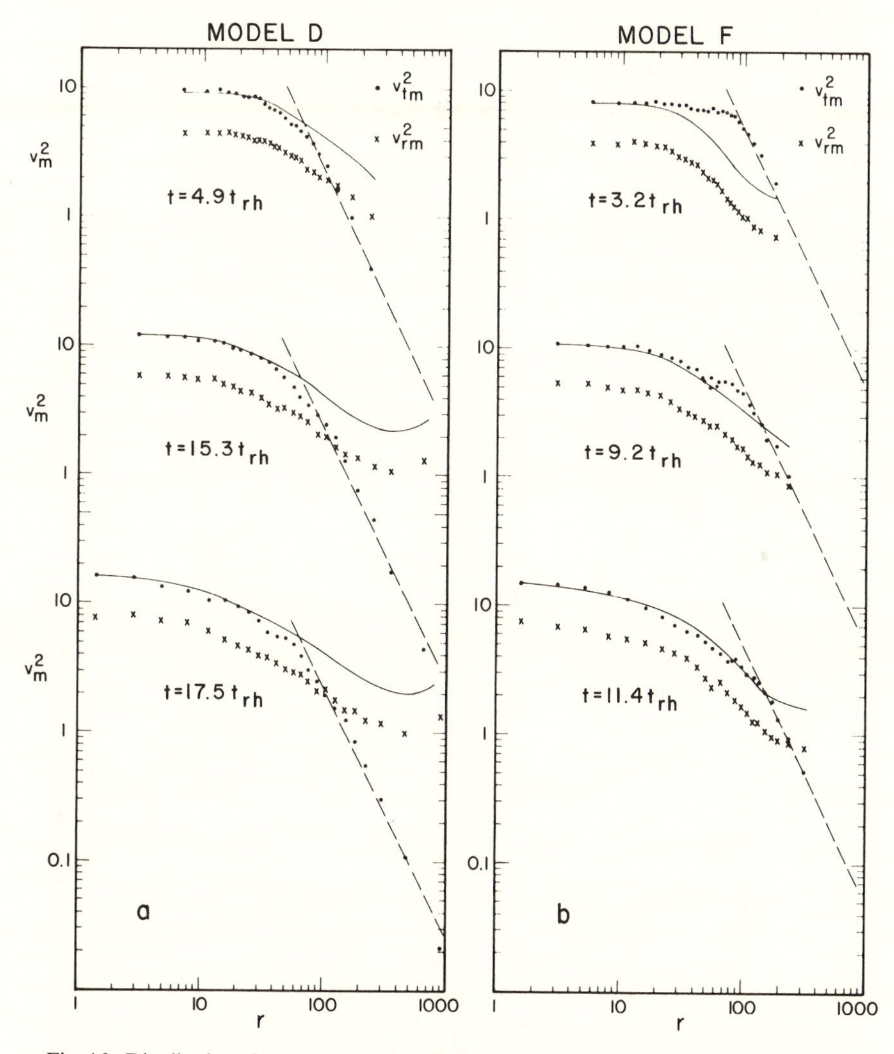

Fig. 4.2. Distribution of Mean Velocities at Different Times. Each plot [3] shows v_{rm}^2 and v_{tm}^2, the mean square velocities in the radial and transverse directions, as functions of r at the same times as in Fig. 4.1. For an isotropic distribution, $v_{tm}^2 = 2v_{rm}^2$, a relation shown by the continuous curve. The dashed line shows v_{tm}^2 varying as $1/r^2$, as expected if the mean square angular momenta were constant throughout the halo.

velocity distribution. In the halo, however, v_{tm} decreases as $1/r$, corresponding to a constant rms angular momentum for the halo stars; v_{rm} decreases more slowly. (The increase at the largest r results from the fact that many of these stars are escaping.) Again we see a clear differentiation of an isolated cluster into an outer radial halo and an inner region which approaches an isothermal sphere more and more closely for regions closer to the center.

The phase-space distribution function $f(E)$ in the central region, where the velocity distribution is isotropic, is shown in Fig. 4.3 for a late stage in the contraction of the system. These results are obtained from a numerical solution of the Fokker-Planck equation, taking into account the anisotropy of the velocity distribution [8]. The indicated $f(E)$ applies not only to all stars near $r \approx 0$ but also to stars in radial orbits, which pass through the central regions. The solid line shows the lowered Maxwellian distribution, which for these isolated clusters is generally too high for E near the escape energy. Similar but less precise results are obtained from the various Monte Carlo models.

The tendency of the cluster towards a Maxwellian velocity distribution function—equation (1-21)—is evident from this figure. Under the influence

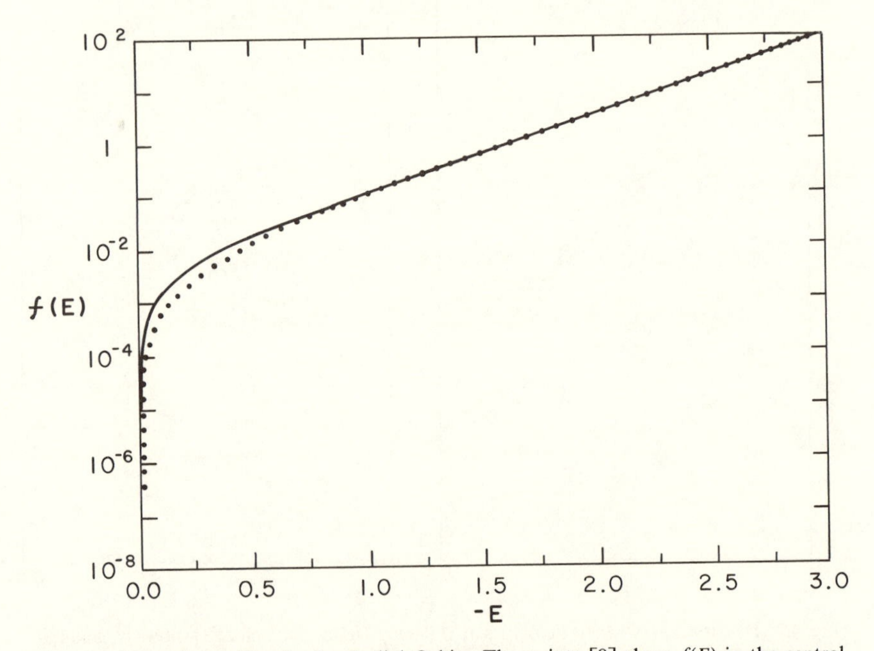

Fig. 4.3. Phase-Space Density for Radial Orbits. The points [8] show $f(E)$ in the central region late in the evolutionary history of the cluster, and apply to radial orbits throughout the system. The solid line represents a lowered Maxwellian distribution.

of mutual encounters, stars tend to diffuse into the region of higher energy, where $f(E)$ is below the Maxwellian value. This process results both in build-up of the halo and in escape of stars.

b. Escape of stars

Within the framework of the small-deflection theory, a star can escape from an isolated cluster only when its energy is already nearly zero, so that in one more passage through the cluster it can acquire a positive energy. Thus escape of stars will not be appreciable until the halo becomes developed and an appreciable number of stars have energies only slightly negative.

This point of view is confirmed by the observed escape rates found in the Monte Carlo calculations for isolated clusters. In Model D the escape probability ξ_e per unit t_{rh} is [3] about 1×10^{-3} during the first half of the cluster's evolutionary history (not including a few stars which escape in the initial collapse and the ensuing oscillations), then rises to about 3×10^{-3}. In Model F, there are no escaping stars during the first two thirds of the time to collapse; during the last third, ξ_e again rises to about 3×10^{-3}. In Model D the finite early escape rate results from the fact that initial collapse produces a halo at the outset. In Model F, however, there is no halo initially, and no escape is possible until a fully developed halo has appeared. Other, more detailed, Monte Carlo calculations [14] show similar increases in ξ_e to about 5×10^{-3} as the halo develops but relatively little change during the final collapse phase. An asymptotic ξ_e of about 4×10^{-3} seems indicated for an isolated cluster with only one mass component.

To understand these model results quantitatively we consider briefly the detailed process by which stars escape as a result of random walk in velocity space. A basic quantity in this process is the rms change of energy in a single orbit; this change, which is a function of E and J, we denote by ε_2. Using the notation defined in equation (2-91), we may write

$$\varepsilon_2^2 \equiv 2 \int_{r_p}^{r_a} \langle (\Delta E)^2 \rangle \frac{dr}{v_r} = P(E,J) \langle (\Delta E)^2 \rangle_{\text{orb}}. \tag{4-9}$$

A similar quantity ε_1, equal to $P(E,J)\langle \Delta E \rangle_{\text{orb}}$, gives the average decrease of E in one orbit. Both ε_1 and ε_2^2, when expressed in dimensionless units (i.e., relative to the mean energy of all cluster stars) are of order $(\ln N)/N$—see equation (4-1). Hence the ratio of ε_1 to ε_2 is very small if N is large and for the present discussion ε_1 can be ignored.

If we follow the evolution of the cluster halo, we see stars in nearly radial orbits diffusing to higher energy in accordance with the Fokker-Planck equation. However, for stars whose energy is comparable with $-\varepsilon_2$, the diffusion approximation is no longer valid, and one must consider the finite change of energy in one orbital passage through the central region. As an approximate way of viewing the phenomenon, we may assume that stars diffuse up to an energy E equal to $-\varepsilon_2$. Stars of greater energy have an appreciable probability of escaping one orbital period later; this probability reaches 50 percent as E rises to zero.

We define the "fringe" of the isolated cluster to comprise the bound stars whose energy lies between $-\varepsilon_2$ and 0. Stars whose energy increases above $-\varepsilon_2$ can escape or they can accumulate in the fringe, where the orbital period approaches infinity as $(-E)^{-3/2}$. In the idealization where the flux of stars diffusing to higher energy through the halo is constant, the halo (excluding the fringe) can reach a steady state, as can the inner regions of the fringe. Most of the stars diffusing upwards in energy will then escape, with energies comparable to the small quantity ε_2. However, as evolution proceeds, a smaller and smaller fraction will accumulate further and further out in the fringe, which cannot reach the steady state found below—see equation (4-10)—since the total number of stars, integrating up to zero energy, would then be infinite—see equation (2-89). Analysis [15] of the time-dependent evolution, on the assumption of a constant diffusive flux toward higher energies, shows that the number of stars which have escaped varies asymptotically as t, since after full halo build-up this flux leads to escape in most cases. In contrast, the total number of stars in the fringe increases only as $t^{1/3}$. As we shall see in chapter 5, these results concerning the fringe are no longer applicable if, as in most actual clusters, the outer radius is limited by an external tidal field.

While the halo of an actual isolated cluster cannot reach an exactly steady state in the course of the cluster's gradual evolution, it can attain a quasi-steady condition in which the conditions of a steady-state model are approximately fulfilled. We now compute the density distribution in the halo corresponding to this condition. One would expect ε_2, the rms change of E in one passage through the cluster core, to change relatively slowly with E, since the velocity of a star as it passes through the central region will be nearly independent of E, if $|E|$ is smaller than the mean binding energy of all cluster stars. If ε_2 is constant, a constant diffusive flux implies a linear increase of $f(E,J)$ with $|E|$. The detailed analysis [15] shows that in fact we have in this case

$$f(E,J) = D_J\left(\frac{1}{2}q + \frac{|E|}{\varepsilon_2}\right), \tag{4-10}$$

where D_J depends on J but not on E, and the quantity q is nearly constant, varying from 1.2 to 1.4 as $|E|/\varepsilon_2$ increases from 0 to 10 or more. The mean energy of the escaping stars, per unit mass, equals $0.58\varepsilon_2$, validating the assumption in §3.1 that evaporating stars have very little kinetic energy at infinity.

The particle density $n(r)$ of bound stars is obtained from equation (1-3), which we write in the form

$$n(r) = \iint f(E,J) 2\pi v_t \, dv_t \, dv_r. \tag{4-11}$$

In equation (1-6) for E we may set $v^2 = v_r^2$, since the transverse velocities are relatively small in the halo. In addition, $\phi(r) \approx -GM/r$, since most of the cluster mass is at smaller radius. Also, in equation (4-10) we neglect the quantity q, which is unimportant for $|E| \gg \varepsilon_2$. The integral over dv_r in equation (4-11) extends from 0 up to $(2GM/r)^{1/2}$, and if we eliminate v_t by use of equation (1-7) we obtain

$$n(r) = \frac{4\pi 2^{1/2}(GM)^{3/2}}{3} \times \frac{1}{r^{3.5}} \times \int D_J \frac{J\,dJ}{\varepsilon_2}. \tag{4-12}$$

In the integral over dJ, D_J is appreciable only for relatively low values of J, since stars diffusing into the halo and escaping are predominantly those in orbits passing through the dense central regions of the cluster and hence have relatively low angular momenta. Since this integral is independent of r, $n(r)$ varies as $r^{-3.5}$, as is shown in Fig. 4.1 and in other Monte Carlo computations [14] for systems with well-developed halos. This exponent of r is comparable to the values noted in §4.2a for a halo formed in the violent relaxation of an initial spherical collapse.

Escape of stars through close encounters will produce a halo with quite different properties from those considered here. The rate of such escape has been computed for the $n = 5$ polytrope (Plummer's model). Although this system has infinite radius, $\rho(r)$ varies as r^{-5} for large r—see equation (1-17). Hence the halo is not fully developed in the sense used here, and ξ_e for the diffusion process resulting from distant encounters is initially nearly zero. When close encounters are considered for this system, the escape rate ξ_e, again defined per time interval t_{rh}, is given by [16]

$$\xi_e = \frac{8.8 \times 10^{-4}}{\ln \Lambda}. \tag{4-13}$$

For $\ln \Lambda$ about equal to 10, equation (4-13) gives an escape rate roughly 0.02 times the value found in the Monte Carlo computations for systems with fully developed halos.

c. Core collapse

Detailed calculations for the standard model show a contraction of the inner core and an expansion of the outer regions. A typical result is shown in

Fig. 4.4, plotted for the same model F portrayed in Figs. 4.1 and 4.2. The solid lines were obtained with the Hénon Monte Carlo method, while the points represent two separate runs with the Princeton method. One unit of time in this figure is about $9t_{rh}$. While these methods differ in many details, one important difference in principle between the two calculations is that

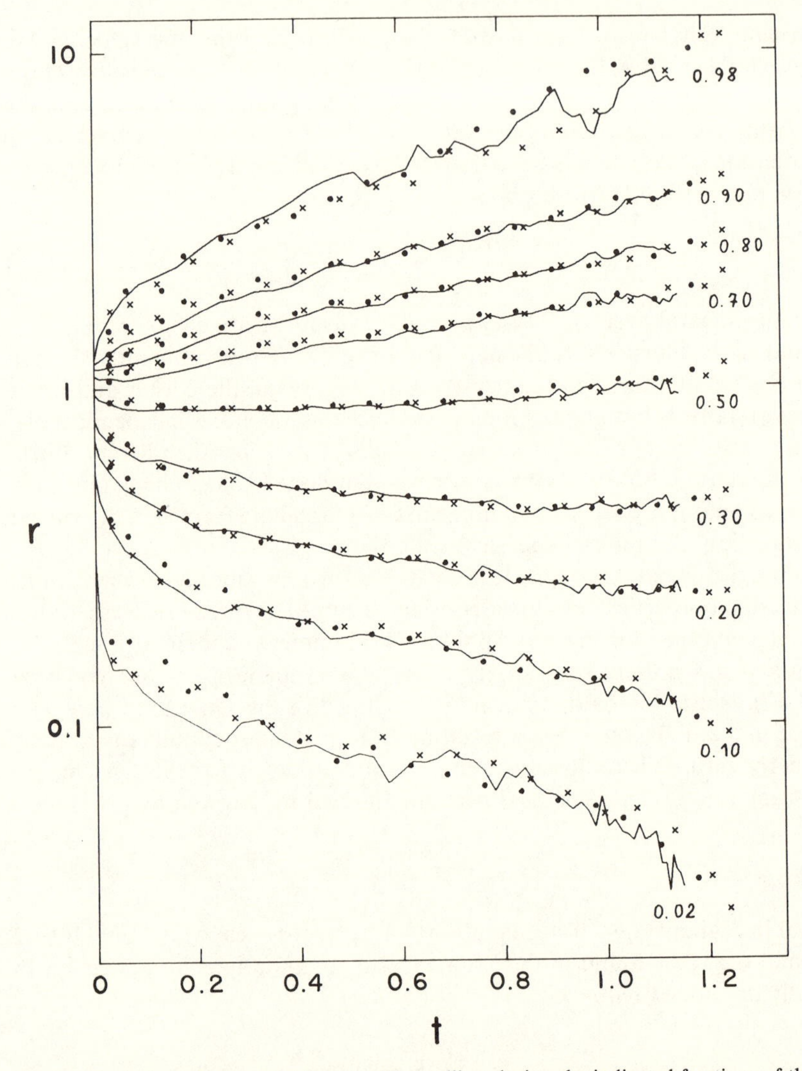

Fig. 4.4. Evolution of Mass Distribution. The radii enclosing the indicated fractions of the total mass are shown [5] as functions of time. The solid curves represent values for Model F—see Figs. 4.1 and 4.2—obtained with the Hénon Monte Carlo method [5]. The points and crosses indicate two different determinations with the corresponding Princeton method [3].

in the latter approach the field stars are assumed to have a Maxwellian distribution, while in the former the stars interact with each other in pairs, and the actual phase-space distribution $f(E,J)$ is included in computing the effects of these interactions. The close agreement evident between these two approaches indicates that the assumption of an isotropic, Maxwellian distribution is a satisfactory approximation for computing the diffusion coefficients in a cluster.

The different curves in Fig. 4.4 show how the radius of the shell containing the indicated fraction of the cluster mass changes with time. The outer regions expand even at the beginning, while the intermediate regions contract. The driving influence during this early phase is the diffusion of stars towards higher energy, where there is a lack of stars relative to the Maxwellian distribution. This diffusion produces halo build-up and escape of stars. Conservation of energy requires a simultaneous contraction of the inner cluster regions.

At a later stage the intermediate regions stop contracting and start to expand. The model computations which extend further into the collapse show this same effect for regions closer and closer to the center, which at progressively later and later times also reverse their contraction and start to expand, just as predicted in the similarity solution for gravothermal collapse—see §3.3b. We summarize here the chief points of comparison between this idealized collapse theory and the model computations. As we shall see, this comparison provides strong confirmation of the theory.

The first quantitative comparison refers to the density ratio $\rho(r)/\rho(0)$ at the point in the cluster where the mean velocity is changing sign, and contraction is changing to expansion. From the rather rough data [3] available in the model calculations, this ratio is about 0.01, as compared with a theoretical prediction of 0.0071.

A second comparison is in the change of the central density with time. If as in §3.1a we introduce the collapse time t_{coll} as the value of t at which the central density ρ_c becomes infinite, then from equations (3-6) and (3-8) we have, for $\zeta = 0.737$,

$$\rho_c^{-0.86} \propto 1 - t/t_{coll}. \tag{4-14}$$

In the detailed model computations the variation of ρ_c with time late in the collapse is in good agreement [17] with equation (4-14). The values found for t_{coll}/t_{rh} depend on initial conditions, varying from 12 to 19.

The relationship at different times between n (or ρ) and v_m at $r = 0$, which we denote by n_c and v_{mc} (equal to the quantity $v_c(t)$ in §3.3b), provides a third comparison. The detailed cluster models give the results shown in Fig. 4.5, where the points represent Monte Carlo results [14], while values obtained from the Fokker-Planck equation [8] are shown by the continuous curve. The variation of n_c as $v_{mc}^{21.2}$, predicted theoretically in equation (3-46), is evidently consistent with the points in this figure, especially at the later times (larger v_{mc}).

Another comparison many be made with the slope of the density distribution function, $\rho(r)$, outside the isothermal core but well inside the inner boundary of the halo. According to the analytic solution, equation (3-45), $d \ln \rho / d \ln r$ equals -2.21 in this region. This result is in very close agreement with the model obtained [9] by solving the Fokker-Planck equation with an isotropic velocity distribution, giving a slope of -2.23. A more crucial if less precise test of the analytical solution as applicable to real clusters is provided by a Cornell Monte Carlo model [17] with $f = f(E,J)$; the slope found for $\ln \rho$ vs. $\ln r$ is about -2.2 over four orders of magnitude change in $\rho(r)$. One may conclude that the self-similar solution for gravothermal collapse provides an excellent approximation for the collapse of the standard model cluster.

Two additional results from the model calculations are of interest. The first refers to the dimensionless central potential X_0, defined by

$$X_0 \equiv -B\phi(0) = -\frac{3\phi(0)}{v_{mc}^2}, \tag{4-15}$$

analogous to the definition of θ in equation (1-25). As before, B is defined in equation (1-22), $\phi(0)$ is the central gravitational potential, with $\phi(\infty) = 0$, and v_{mc} is the rms three-dimensional velocity dispersion at $r = 0$. In Plummer's model, $X_0 = 6$ according to equation (1-19). In the evolutionary calculations which start from this model, X_0 increases steadily as the cluster first contracts, with a more rapid increase when $X_0 \geqslant 9$, apparently indicating the onset of gravothermal collapse. The observational tests of the gravothermal collapse theory which are given above all refer to properties of model clusters for $X_0 \geqslant 9$. Many of the model calculations do not extend beyond the range between 10 and 12 for X_0; however, in a few models X_0 reaches 14, as equation (3-45) for $\rho(r)$ becomes valid over a larger and larger range of r. In the self-similar solution of §3.3b, this range is essentially infinite and $X_0 = 15.3$.

A second significant quantity is the collapse rate ξ_c which appears in equation (3-29) and which measures the rate of increase of $\rho_c(t)$. In the models ξ_c is initially about 0.2; after a decrease early in the evolution, ξ_c starts to flatten out when X_0 reaches 9, with only a slight further decrease beyond $X_0 = 10$. Model clusters obtained with numerical solution of the Fokker-Planck equation [8,9] show the following asymptotic results:

$$\xi_c = \begin{cases} 0.006 & \text{for } f = f(E,J) \\ 0.0036 & \text{for } f = f(E). \end{cases} \tag{4-16}$$

The value given for the anisotropic f is the less accurate of the two.

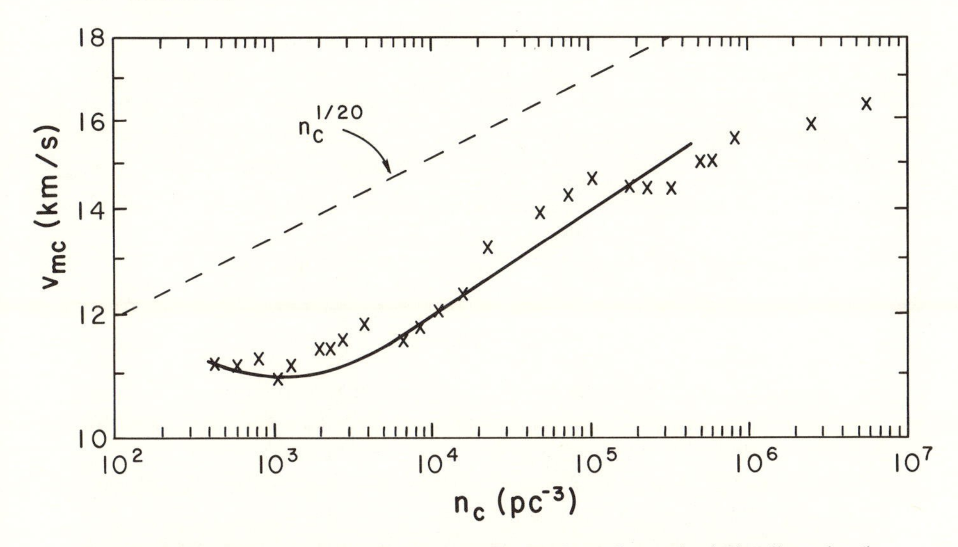

Fig. 4.5. Relationship of v_m and n at Cluster Center. Values of central velocity dispersion (in km/s) are plotted [14] against central density (in stars per cubic parsec) late in the evolution of a collapsing cluster core. The points are obtained with the Cornell Monte Carlo program [14], while the solid line results from a numerical solution of the Fokker-Planck equation [8] for an anisotropic velocity distribution. The dashed line shows for comparison a variation of v_{mc} as $n_c^{1/20}$.

Confirmation of the higher value of ξ_c for the more realistic anisotropic case is provided by a Cornell Monte Carlo model [17]; the plotted data which confirm equation (4-14) may be used to show that for $X_0 \geqslant 9$, $\xi_c = 6 \times 10^{-3}$, if equation (2-61) rather than (2-75) is used for t_{rc}, and if $\ln \Lambda = 11.7$ in t_{rc} as well as in t_{rh}. In fact a decrease of $\ln \Lambda$ with decreasing N_c was assumed in this model; if in t_{rc} we take $\ln \Lambda = 8.3$, a rough average of the values assumed during gravothermal collapse, ξ_c is increased by a factor 1.4 to 8×10^{-3}. In view of the statistical fluctuations inherent in Monte Carlo computations, we adopt the value 6×10^{-3} in equation (4-16) as perhaps the more reliable result for the anisotropic case.

From these results, together with equation (3-47), we obtain for the number of central relaxation times remaining until complete collapse

$$\frac{\tau}{t_{rc}} = \begin{cases} 190 & \text{for } f = f(E,J), \\ 320 & \text{for } f = f(E). \end{cases} \tag{4-17}$$

In this equation t_{rc} is evaluated, of course, at the time $t = t_{\text{coll}} - \tau$.

It is perhaps surprising that the models based on an isotropic $f(E)$ should agree so well with the more general models in which the dependence of f on J is properly considered. The temperature gradients associated with the gravothermal instability would be expected to produce anisotropic velocity distributions in the inner, contracting regions; heat can be carried directly from the inner to outer regions by stars in radial orbits. Such anisotropies at small r are in fact revealed in the models late in the collapse [17,18] but do not seem to have any large effect on the structure of the collapsing core. However, the mechanism of heat transfer producing these anisotropies may be responsible for the increased rate of collapse shown in equation (4-16) for the anisotropic model in comparison with the isotropic one.

d. Mass stratification

The numerical models with several mass components all show rapid stratification of stars by mass, with the heaviest stars in particular tending to concentrate in the cluster core during a time not much longer than t_{rh}. Evidence on the possible presence of the mass stratification instability, discussed in §3.4, is more difficult to obtain, since the core collapses in any case, in large part because of the gravothermal instability. However, the differences of kinetic energy which appear during the core contraction might be expected to have some relationship with condition (3-55).

The rate at which mass stratification occurs in a three-component cluster is shown in Fig. 4.6, where the radii containing half the mass of each component are plotted [19] as a function of t/t_{rh}. The masses of the three components are in the ratios: 1, 2.5, 6.25, with total masses in the ratios 1/2, 1/3 and 1/6. In these Princeton computations, Maxwellian velocity distributions were assumed for each of the three mass components as field particles, and velocity perturbations were computed for each test star in interactions with all three groups of field stars, again using equations (2-52) to (2-54). The reference relaxation time, t_{rh}, was determined again from equation (2-63), with m set equal to the mean mass per star, equal to M/N, the ratio of total mass to total star number; this is a useful first approximation, but can give errors of two or more in some situations if equipartition is reached. The concept of a reference relaxation time must be used with caution when components of different masses are present, especially when their relative distribution changes with time. Subject to this approximation, the presence of several mass components evidently leads within a time of one or two times t_{rh} to the concentration of the more massive stars within the central region.

After the heavier stars become concentrated near the center, the contraction of the cluster proceeds in much the same way as in a single-component system. To demonstrate this conclusion, Fig. 4.7 plots for a two-component cluster, with a stellar mass ratio of 2 to 1, the values [10] of the dimensionless contraction rate ξ_c, defined in equation (3-29). The ratio of

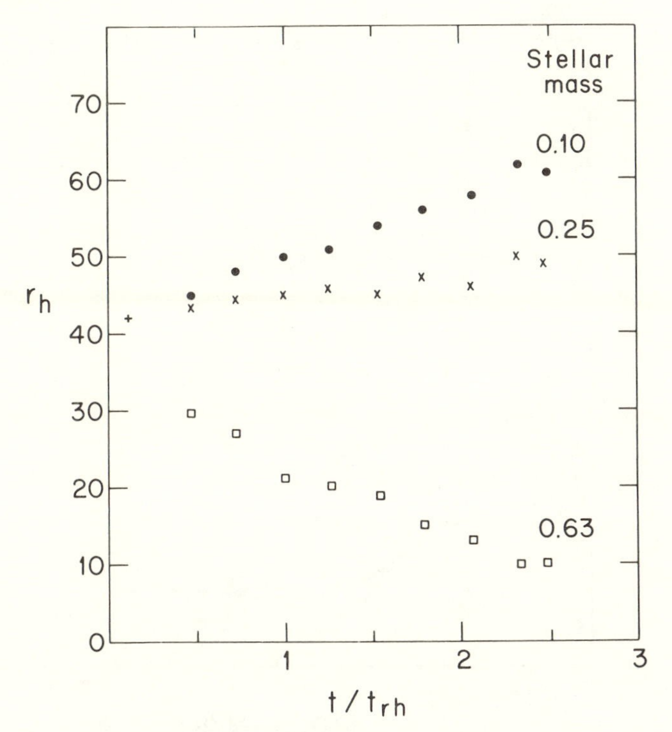

Fig. 4.6. Rate of Mass Stratification. For a cluster containing three mass components the change of the median radius r_h for each component is shown [19] as a function of time, obtained from a Monte Carlo model. The fraction of cluster mass in the heaviest stars was 1/6, with 1/3 and 1/2 in the intermediate and lightest groups, respectively. The symbol at lowest t shows the initial value of r_h containing half the total mass.

central densities in the two components is also plotted. In this model, based on a numerical solution of the Fokker-Planck equation with an assumed isotropic velocity distribution, the ratio M_2/M_1 of total masses is 1/99. The curves for ξ_{c1} and ξ_{c2} show the values of ξ_c obtained for each component separately, ignoring the other component in the computation of central density, ρ_c, and central relaxation time, t_{rc}. The horizontal scale is the dimensionless central potential, with R the scale factor in the Plummer's model assumed at the start—see equation (1-19). (All values of ξ_c have been multiplied by 3/4 to convert from t_R to t_r in equation (3-29).)

At the beginning both components have the same distribution, giving ρ_{c2}/ρ_{c1} equal initially to M_2/M_1. As the cluster evolves, the ratio ρ_{c2}/ρ_{c1} increases, and equals 1 for $\phi(0)/(GM/R) = -2.8$. Fig. 4.7 shows that for later times, ξ_{c2} approaches the value of ξ_c for a system with only one mass

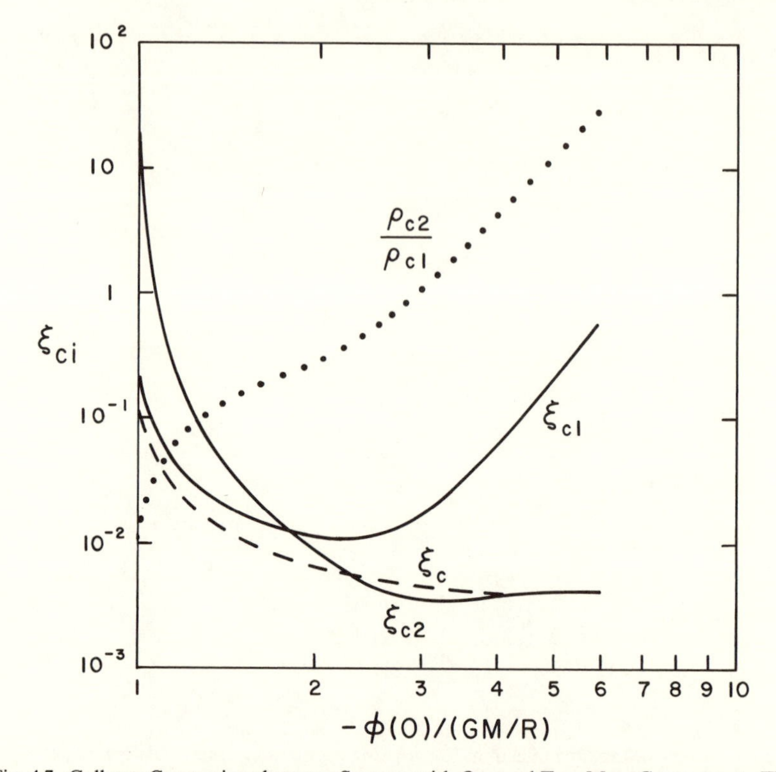

Fig. 4.7. Collapse Comparison between Systems with One and Two Mass Components. The two solid curves represent $t_{ri}\, d \ln \rho_{ci}/dt$ for the lighter (ξ_{c1}) and the heavier component (ξ_{c2}) in a two-component system; ρ_{ci} is the central density of component i, and t_{ri} is the central relaxation time for component i in interactions with similar stars. These values were obtained [10] from a numerical solution of the Fokker-Planck equation. The dashed curve shows the corresponding central density growth rate ξ_c for a single-component system; the dotted curve shows the ratio of central densities, ρ_{c2}/ρ_{c1}. The horizontal scale is the dimensionless central potential; R is the scale factor for the Plummer model cluster assumed initially.

component; the two curves are indistinguishable for $\rho_{c2}/\rho_{c1} \geqslant 5$, approaching the same value 3.6×10^{-3} given in equation (4-16) for a single-component cluster with an isotropic velocity distribution. This evidence indicates that the late collapse of a two-component isolated cluster results mostly from the gravothermal instability, since we have seen above that this effect is clearly dominant in the corresponding evolutionary stage for a one-component system.

The same conclusion is also indicated by the radial density gradient shown in another such model [18]. Late in the cluster evolution,

$d \ln \rho_2/d \ln r$ for the heavier component is about equal to the same value, -2.23, obtained in isotropic models with only one component.

The effects associated with the mass stratification instability (discussed in §3.4) are presumably responsible for the difference of kinetic energies between stars of the two masses, as observed in these same numerical solutions of the Fokker-Planck equation [10]. For small M_2/M_1 and at early times the stars of greater mass m_2 rapidly approach equipartition of energy near the cluster core with the lighter stars, of mass m_1. The kinetic energies per star diverge only late in the collapse phase, when the central density ρ_{c2} has exceeded ρ_{c1}. If $M_2 = M_1$, ρ_{c2} exceeds ρ_{c1} very soon in the course of evolution, and equipartition is never established even at the center. In a general way departures from equipartition seem to be correlated with the quantity χ defined in equation (3-54).

The primary effect of mass stratification on cluster evolution is to shorten the time interval from the birth of the cluster as a stellar system until the rapid collapse of the central core. The relatively rapid development of a central core composed of the more massive stars leads to an early onset of the gravothermal instability. Computations with many mass components, simulating a continuous distribution of masses between m_{min} and m_{max}, show similar results [20]. For reasonable assumptions about the mass distribution ($m_{\text{max}}/m_{\text{min}} = 10$, and $-d \ln N(m)/d \ln m$ between 2 and 4), t_{coll} for a multi-component cluster (its age when ρ_c approaches infinity) is between 2 and 4 times t_{rh}, as compared to a typical value of $15t_{rh}$ for a one-component system.

REFERENCES

1. S. J. Aarseth and M. Lecar, *Ann. Review Astron. Astrophys.*, **13**, 1, 1975.
2. *Gravitational N-Body Problem*, IAU Colloq. No. 10, ed. M. Lecar (Reidel, Dordrecht), 1972.
3. L. Spitzer and T. X. Thuan, *Ap. J.*, **175**, 31, 1972.
4. M. Hénon, *Gravitational N-Body Problem*, IAU Colloq. No. 10, ed. M. Lecar (Reidel, Dordrecht), 1972, p. 406; *Astroph. and Space Sci.*, **14**, 151, 1971.
5. M. Hénon, *Dynamical Structure and Evolution of Stellar Systems*, ed. L. Martinet and M. Mayor (Geneva Obs.) 1973, p. 183.
6. S. L. Shapiro and A. B. Marchant, *Ap. J.*, **225**, 603, 1978; see Appendix.
7. A. B. Marchant and S. L. Shapiro, *Ap. J.*, **234**, 317, 1979.
8. H. Cohn, *Ap. J.*, **234**, 1036, 1979.
9. H. Cohn, *ibid.* **242**, 765, 1980.
10. S. Inagaki and P. Wiyanto, *Publ. Astr. Soc. Japan*, **36**, 391, 1984.

11. T. S. Statler, J. P. Ostriker and H. N. Cohn, *Ap. J.*, **316**, 626, 1987.
12. M. Hénon, *Symposium on Computer Simulation of Plasma and Many-Body Problems*, NASA SP-153, 1967, p. 349.
13. T. A. McGlynn, *Ap. J.*, **281**, 13, 1984.
14. A. B. Marchant and S. L. Shapiro, *Ap. J.*, **239**, 685, 1980.
15. L. Spitzer and S. L. Shapiro, *Ap. J.*, **173**, 529, 1972.
16. M. Hénon, *Ann. d'Astroph.*, **23**, 668, 1960.
17. M. J. Duncan and S. L. Shapiro, *Ap. J.*, **253**, 921, 1982.
18. H. Cohn, *Dynamics of Star Clusters*, IAU Symp. No. 113, ed. J. Goodman and P. Hut (Reidel, Dordrecht) 1985, p. 161.
19. L. Spitzer and J. M. Shull, *Ap. J.*, **201**, 773, 1975.
20. S. Inagaki and W. C. Saslaw, *Ap. J.*, **292**, 339, 1985.

5

Effects of External Fields

A conspicuous lack of realism in the standard model is its neglect of the galactic gravitational field. The tidal force associated with this field makes it possible for individual stars to escape from the cluster if their energy suffices to take them beyond some critical distance from the cluster center. In addition, variations in the galactic field, as experienced by a cluster moving about the Galaxy, provide a time-dependent gravitational potential, when expressed relative to the center of the cluster, and can produce important changes in the random kinetic energies of some cluster stars.

These effects are discussed in the following two sections. The first of these deals with the tidal effects of the galactic field, assuming spherical symmetry for this field and ignoring the heating associated with varying R_G, the distance of the cluster from the galactic center. Section 5.2 examines the heating effects associated with time-dependent fields, first those experienced when a cluster passes through the galactic disc, and then those resulting from cluster passage near the galactic center, where R_G has its minimum value at perigalacticon.

5.1 TIDAL GALACTIC FIELD

To the extent that the Galaxy is treated as a spherical mass, around which the globular cluster moves in a circular orbit, the equations of motion permit an exact energy integral, and the conditions for possible escape of stars are relatively straightforward. If the cluster orbit is eccentric, analysis gives only approximate results, with no exact necessary condition for escape. In any case detailed computations of stellar orbits are required to indicate when escape actually occurs. These topics are discussed in the first subsection, while the second applies these results, and especially the computed value of the effective escape radius r_t, to the rate of escape ξ_e for cluster stars.

a. Exact results for a steady field

We have already given in §1.2b an approximate analysis of the galactic tidal field experienced by a globular cluster, on the artificial assumption that the centers of the cluster and the Galaxy are fixed relative to each other. On the more realistic assumption that the cluster is in a circular orbit around the galactic center, again at a distance R_G, the motion of a single cluster star, with negligible mass, is a familiar problem in the motion of three bodies. If we assume first that all three masses may be regarded as points, the effective potential in the rotating coordinate system fixed with respect to the cluster and the Galaxy may be written

$$-\phi(\mathbf{r}) = \frac{GM_G}{r_G} + \frac{GM_C}{r_C} + \frac{\omega^2}{2}(x_0^2 + y_0^2), \tag{5-1}$$

where r_G and r_C are the distances from the star to the Galaxy and the cluster, respectively, whose masses are M_G and M_C. The quantities x_0 and y_0 are components of $\mathbf{r}$ measured from the center of gravity of the cluster-Galaxy system; $\mathbf{x}_0$ is measured parallel to $\mathbf{R}_G$, with $\mathbf{y}_0$ perpendicular to $\mathbf{x}_0$ and in the cluster orbital plane. The last term in equation (5-1) is the familiar centrifugal potential in a frame of reference uniformly rotating with angular velocity ω. If equation (5-1) is inserted into equation (1-6) for the total energy, with $\mathbf{v}$ measured in the rotating frame, one obtains "Jacobi's integral" for this restricted problem; the equipotential surfaces defined by equation (5-1) are called "surfaces of zero velocity."

If we measure x and y from the cluster center, with x (like x_0) positive in the direction of the galactic center, and if we define μ as the mass ratio M_C/M_G, we have

$$x = \frac{R_G}{1 + \mu} + x_0; \tag{5-2}$$

y_0 and z_0 equal y and z, respectively. The angular velocity of the rotating frame is given by

$$\omega^2 = \frac{GM_G(1 + \mu)}{R_G^3}. \tag{5-3}$$

If we expand $1/r_G$ in powers of x/R_G, y/R_G and z/R_G, retaining terms to second order, and neglecting terms of order μ, we find that equation (5-1) may be written

$$-\phi(\mathbf{r}) = \frac{1}{2}\frac{GM_G}{R_G^3}(3x^2 - z^2) + \frac{GM_C}{r_C} + A. \tag{5-4}$$

Along the x axis the centrifugal potential has the same sign as the tidal potential and is half as great, while along the y axis the two potentials make equal and opposite contributions. Since the zero point of the potential is arbitrary, we have replaced the constant term $3GM_G/2R_G$ obtained from equation (5-1) with the constant A. In the following discussion we set $A = 0$.

We let x_e be the value of x at which $\partial\phi/\partial x = 0$, for $y = z = 0$; $\phi(x_e)$ is the potential at this point on the x axis. The potential surface passing through this point is the Roche surface, familiar from the theory of evolving binary systems. With straightforward calculations based on equation (5-4), we find

$$x_e^3 = \frac{1}{3}\frac{M_C}{M_G} R_G^3, \tag{5-5}$$

and

$$\phi(x_e) = -\frac{3GM_C}{2x_e}. \tag{5-6}$$

Equation (5-5) is the first term in a series expansion [1] of x_e/R_G in powers of $\mu^{1/3}$. Consideration of the rotating frame of reference, as the cluster circles around the galactic center, has reduced x_e by a factor $(2/3)^{1/3} = 0.87$ times the value obtained in equation (1-33).

Fig. 5.1 shows the intersection of several equipotential surfaces with the xy plane; for each curve the quantity $\psi \equiv \phi(\mathbf{r})/\phi(x_e)$ is indicated, where

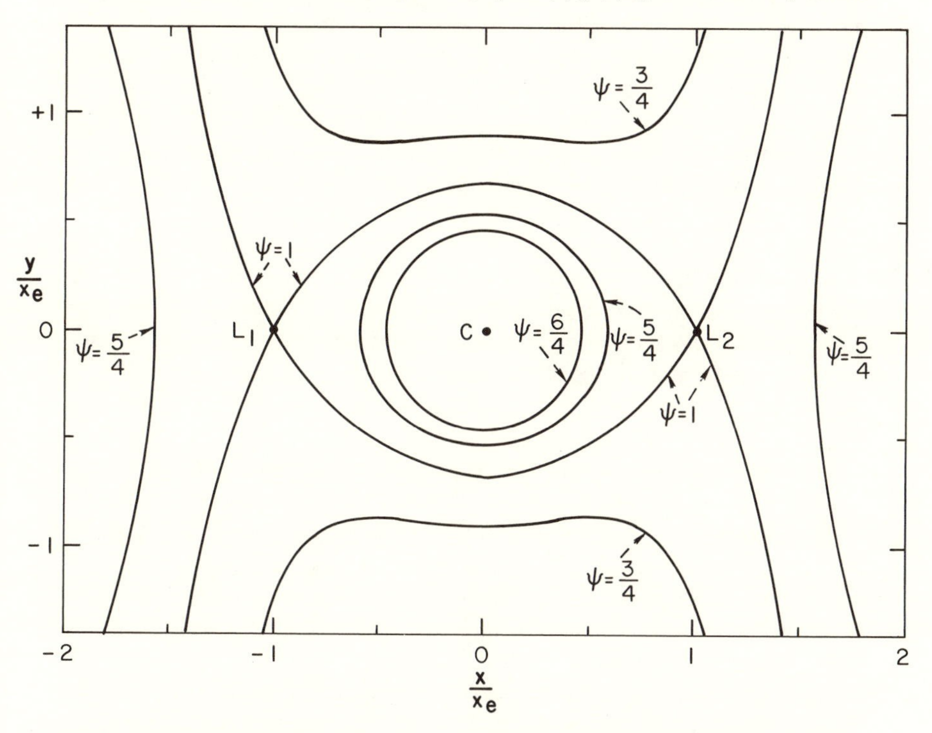

Fig. 5.1. Equipotential Surfaces with Steady Tidal Field. The contours represent constant values of the total dimensionless potential ψ (gravitational plus centrifugal) in the xy plane through a cluster subject to the tidal force of the galactic center (in the $+x$ direction). The cluster revolves about the galactic center in a circular orbit. The two Lagrangian points, L_1 and L_2, where the net force vanishes, are at a distance x_e from the cluster center at C.

$\phi(x_e)$ is given in equation (5-6). Such surfaces in general all have reflection symmetry in the xy and xz planes. In addition, for the present case, with $\mu \equiv M_C/M_G \ll 1$, this same symmetry exists with respect to the yz plane, if terms of order $(r/R_G)^3$ and higher are ignored in ϕ. Thus the two Lagrangian points, L_1 and L_2, are at distances $-x_e$ and $+x_e$ from the cluster center. These are evidently saddle points, with ψ a minimum in the x direction and a maximum in the transverse y and z directions. The curve $\psi = 1$ through these two points represents the intersection of the critical Roche surface with the xy plane.

It is clear that in the absence of encounters any star with energy per unit mass less than $\phi(\mathbf{x}_e)$ cannot escape, since it cannot reach the Roche surface. For stars with $E > \phi(\mathbf{x}_e)$, escape is energetically possible, since the equipotential surfaces for such energies extend far around the Galaxy. Escape of all such stars is by no means assured, since the relationship of particle orbits to the equipotential surfaces is somewhat indirect, especially in view of the Coriolis acceleration $\boldsymbol{\omega} \times (\boldsymbol{\omega} \times \mathbf{v})$ characteristic of rotating systems. While stars whose orbits take them beyond the Roche surface are likely to escape in time, possibly after many orbital periods, the density distribution of the cluster may be expected to extend outside this critical surface. This extended density profile will depend primarily on shock heating, discussed below. However, as a first approximation the presence of such bound stars with $E > \phi(x_e)$ is usually ignored in theoretical models.

Such models are based also on the simplifying assumption of spherical symmetry, with the potential set equal to $-GM_C/r$. Equation (5-6) then indicates that the tidal cut-off radius r_t at which $\phi = \phi(x_e)$ and beyond which the density vanishes is given by

$$r_t = 2x_e/3, \tag{5-7}$$

where x_e is given by equation (5-5). The factor 2/3 appearing here is evident also in Fig. 5.1, where the distance to the critical potential in the y direction is two thirds of x_e; in the y direction the effective force is entirely produced by the cluster alone. Since r_t for observed clusters is generally determined by extrapolation of $\rho(r)$ in regions where the equipotentials are still circular, the values of r_t found should in principle include the 2/3 factor in equation (5-7).

Equation (5-4) may readily be generalized to take into account the galactic mass distribution, provided spherical symmetry is assumed. For the cluster this refinement is unnecessary, since in the regions where the galactic tidal force is significant the number of stars per unit volume is relatively low, and the cluster potential can be regarded as that of a point source. In the Galaxy, however, there is evidence that the density of the gravitating mass varies as $1/R^2$, and that hence $M_G(R)$, the galactic mass within the radius R, increases linearly with R. If the resultant galactic potential is introduced into equation (5-1), the tidal force in the x direction is halved. The tidal forces in the y and z directions are unchanged, since these depend not on $\partial\phi/\partial r_G$ but on the change in direction of $\mathbf{r}_G$ as y and z change. Thus the one change in equation (5-4) is that the term $3x^2$ becomes $2x^2$; in equation (5-5) the factor 1/3 is replaced by 1/2. Equation (5-6) is unaffected.

These results are all substantially modified if, as generally assumed, the cluster is in a somewhat eccentric orbit around the galactic center. The tidal force from the Galaxy is then greatest at perigalacticon, where the separation of the cluster center from the galactic center equals R_p. If a point mass at the galactic center is again assumed, R_p is related to the eccentricity e and the semi-major axis a of the cluster's orbit by the usual equation

$$R_p = a(1 - e). \tag{5-8}$$

Since the potential is now time-dependent, the angular velocity ω of the cluster in its orbit varies with time. At perigalacticon ω may be determined from equation (5-8) together with the orbital energy of the cluster per unit mass, which equals $-GM_G/2a$. If V_p denotes the velocity of the cluster at perigalacticon, we obtain

$$\omega^2 = \frac{V_p^2}{R_p^2} = \frac{2E + 2GM_G/R_p}{R_p^2} = (1 + e)\frac{GM_G}{R_p^3}. \tag{5-9}$$

Since ω and R_G are both varying with time, there is no energy integral for individual stars in the cluster, and equation (5-1) is not applicable; no necessary condition can be given for escape of cluster stars. To obtain an approximate value for the tidal cut-off we consider the acceleration of a star which at perigalacticon is located on the x axis, joining the centers of cluster and Galaxy. We define x_e as the distance of the star from the cluster center such that the stellar acceleration in the x direction equals the corresponding acceleration of the cluster center toward the Galaxy. Stars closer than x_e to the cluster center will tend to draw nearer to the cluster, while those at larger x will tend to be drawn away. If we equate $\omega^2 R_p - GM_G/R_p^2$ for the cluster to the corresponding acceleration of the cluster star, at the distance $R_p - x_e$ from the galactic center, use of equation (5-9) gives [2]

$$x_e^3 = \frac{1}{3 + e}\frac{M_C}{M_G}R_p^3, \tag{5-10}$$

in agreement with equation (5-5) for $e = 0$, $R_p = R_G$. According to equation (5-8), for moderate eccentricity R_p is substantially less than a; hence equation (5-10) can give a much smaller value of x_e than is found with R_G set equal to a in equation (5-5). Equation (5-10) gives what is sometimes referred to as the King tidal radius [2]; i.e., the value of r_t beyond which the cluster density is assumed to vanish.

The use of equation (5-10) to determine r_t when e is finite is evidently very approximate. Hence we shall ignore here the various corrections discussed above; i.e., the linear variation of M_G with R_G, which increases r_t, and the use of equation (5-7), which decreases it. Instead we rely on numerical computations to give the value of r_t for stellar escape from a cluster which moves around the Galaxy in an eccentric orbit. Such computations do in fact confirm that in a general way escape is likely if the maximum distance of a star from the cluster center exceeds the King tidal radius, as given in equation (5-10). Dynamical integrations of stellar orbits have been carried out [3] in a cluster revolving about a point mass at the galactic center in an orbit with $a = 6$ kpc and $e = 2/3$, giving $R_p =$ 2 kpc. Values of M_C and M_G were set equal to $7 \times 10^5 M_\odot$ and $10^{10} M_\odot$, respectively, giving $x_e = 53$ pc from equation (5-10). The cluster halo was assumed to extend initially out to 75 pc; stars were considered whose orbits in the cluster had apocenters at different distances and a variety of eccentricities and inclinations.

The computations showed that after one cluster passage through perigalacticon, stars in eccentric orbits ($e \geqslant 0.5$) tended to escape if their apocluster radius, r_a, exceeded 50 pc. For more nearly circular orbits of zero inclination the results differed markedly between direct and retrograde orbits, with the former escaping for r_a greater than some 40 pc, with the latter remaining bound even for r_a as great as 80 pc. Since the angular velocity of a star in a circular orbit at 50 pc is 5×10^{-15}/s under these conditions, somewhat exceeding $V_p/R_p = 3 \times 10^{-15}$/s for the cluster at perigalacticon, this large difference between direct and retrograde orbits is to be expected. For circular high inclination orbits ($i \geqslant 60°$) escape is quite likely (5 out of 8 cases) for apocluster distance as great as 55 pc (with a marked difference between direct and retrograde orbits), but is not found at 40 pc. The situation is complicated by the heating effect produced by this time-dependent potential, an effect discussed in §5.2c, and no precise comparison between theory and simulation is possible.

It is difficult to apply equation (5-10) to actual clusters since the perigalacticon distance, R_p, is unknown. A statistical approach has been used [4] to examine whether the distribution of the observed r_t values for actual clusters is consistent with an isotropic distribution of cluster velocities with respect to the galactic center. The results indicate an absence of the relatively low values of r_t and high values of the mean cluster density, $3M_C/4\pi r_t^3$, that would be expected in clusters with predominantly radial orbits and consequently low values of perigalacticon radius, R_p. If the observed tidal radii are assumed to be correct, either most globular clusters have sufficient angular momenta to keep them out of the central regions of the Galaxy, or else close perigalacticon passages are less effective than assumed in stripping a cluster.

b. Evaporation from modified standard model

The observations described in §1.1 and the theory presented above show clearly that globular clusters near a galaxy will be limited by the galactic tidal field. This effect can be taken into account approximately by modifying the standard model to include a tidal cut-off at a radius r_t, set equal to x_e in equation (5-10). Computations based on the assumption of immediate escape of all stars with $r > r_t$ should indicate in a general way how a tidal cut-off influences the rate of escape.

Several models of such truncated spherical clusters have been computed with Monte Carlo techniques. Since these computations did not extend very far into the collapse phase, the variation of the bound cluster mass M with time was not very large, and the variation of r_t with M was ignored. When a single mass component was assumed [5], the value of the escape probability ξ_e per unit t_{rh}—see equation (3-3)—was found to increase from 0.015 to 0.05 as r_h/r_t was varied from 0.11 to 0.34, covering the range for realistic King models (see §1.2b). For the Hénon tidally truncated model [6], in which $r_h/r_t = 0.145$, the escape probability per time interval t_{rh} equals 0.045—see equation (3-27)—if t_{rh} is computed from equation (2-63); this value of ξ_e is about twice as great as indicated by the Monte Carlo models, presumably because of the various differences between these idealized clusters.

The variation of ξ_e with stellar mass, when different mass components are present in the cluster, is also shown in Monte Carlo models [7]. For models with half the cluster mass in the heaviest component, the relative escape probabilities for stars of different masses are shown in Table 5.1; results are

TABLE 5.1
Relative Rate of Escape for Different Mass Components

Relative stellar mass	1	0.4	0.16
Relative escape rate, early	1	1.6	2.4
Relative escape rate, late	1	2.3	3.3
Relative escape rate, Hénon model	1	2.4	3.1

given separately for the first third (early) and the last third (late) of the time interval covered by the computations. The escape probabilities, which are about twice as great during the late interval as during the early one, have been normalized to the values observed for the heaviest stars. The agreement with the values for the Hénon model, taken from Table 3.1 (p. 62), is remarkably close for the later phase, when mass stratification has developed; in the self-similar Fokker-Planck model virtually all the mass was assumed to reside in the heaviest stars.

If the evaporation probability, ξ_e, is set equal to 4.5×10^{-2} per time interval t_{rh}, as found by Hénon for the dominant stars, and is then multiplied by 3 to obtain ξ_e for the lightest stars ($m/m_f \leqslant 0.2$), we find that about 90 percent of the lightest stars will escape in at most $17t_{rh}$. According to Fig. 1.3 (p. 6), in about two-fifths of the observed globular clusters the number of the lightest stars will decrease by at least an order of magnitude during a life of 17×10^9 years.

As we have seen in §3.2, a linear decrease of $M(t)$ with time is a characteristic of tidally truncated models which are contracting homologously. It is significant that a nearly linear such decrease was found in a reasonably realistic model cluster [8], in which most relevant physical effects, such as formation of binaries, stellar mass loss and mass stratification, were all considered. The value of ξ_e obtained had a nearly constant value of about 0.05, in agreement with equation (3-27) for the Hénon model, even though the contraction was by no means homologous; for example, r_h/r_t rose from an initial value of 0.12 to about 0.3 at 2×10^{10} years.

5.2 TIME-DEPENDENT FIELDS

When $\nabla\phi$, the gradient of the gravitational potential, changes with time in the region where a particle is moving, the energy E per unit mass of the particle will not be an integral of the motion. In the general case the energy of any one star can be increased or decreased by such a time-dependent perturbation; in the idealized cases considered here, the average energy of the particles tends to increase.

The energy change can be very small if the duration $t_{\rm per}$ of the transient perturbation is much greater than the orbital period P of the star within the cluster. In this case, the integrals of $v_i\,dx_i$ around an orbit (the action integrals) are adiabatic invariants, and tend to be constant, as noted above in §2.3. When these integrals are constant, the orbit after the perturbation disappears will be identical with the orbit beforehand, and in particular the energy will be the same. However, this constancy is not exact. We let β be a dimensionless constant proportional to the ratio $t_{\rm per}/P$. As noted in §2.3, when β increases, the change of the action integral resulting from the perturbation approaches zero in simple cases more rapidly than any power of $1/\beta$ (as, for example, does an exponential function of $-\beta$). When the perturbation results from the evolutionary change of the cluster, β is of order $t_{rh} \times v_m/r_h$, which usually exceeds at least 10^2—see §1.1. With so large a value of β, the potential changes very slowly compared to the orbital period, and for any one star the change of the radial action integral resulting from $\partial\phi/\partial t$ should be very small indeed. In the cases we consider here β is not so large and may be small compared to unity. In such cases a transient perturbation may leave cluster stars in states of appreciably different energy from their initial values.

In the first subsection below we discuss the gravitational perturbation resulting from passage of a globular cluster through the galactic disc. The perturbing potential may be accurately represented by a one-dimensional model in which the potential ϕ_d of the galactic disc is a function only of the vertical coordinate Z and is constant in time. As seen in a reference frame travelling with the cluster, $d\phi_d/dZ$ changes with time, providing a transient perturbation. Two other types of gravitational perturbations are considered subsequently—one produced by an encounter between two clusters, the other produced by the fluctuating potential of the galactic center, as seen by a cluster moving in an eccentric orbit around this central mass. All these perturbations modify the random kinetic energy of stars in the outer regions of the cluster, accelerating escape of stars and the evolution of the cluster.

a. Theory of heating by compressive gravitational shocks

The gravitational potential of the galactic disc is given by equation (1-5), which here becomes

$$\frac{d^2\phi_d}{dZ^2} = 4\pi G\rho_d(Z), \tag{5-11}$$

where we denote by Z the distance above the galactic plane; ρ_d is the density of gravitating matter within the disc and is assumed to be a function of Z only. The gravitational acceleration, $g = -d\phi_d/dZ$, is zero in the plane and its magnitude rises to a nearly constant value for $|Z|$ sufficiently large that ρ_d has fallen to a very low value.

As a cluster passes through this disc, the cluster center, at a distance Z_C from the midplane, will be subject to an acceleration $g(Z_C)$, while a cluster star, at a vertical distance Z from the midplane, will be subject to a different acceleration, $g(Z)$. If we denote $Z - Z_C$ by z, the distance of the star from the cluster center, measured perpendicular to the plane, and denote dz/dt by v_z, then the galactic disc will produce a relative acceleration which we denote by $(dv_z/dt)_d$, given by

$$\left(\frac{dv_z}{dt}\right)_d = g(Z) - g(Z_C) = z\frac{dg}{dZ}(Z_C), \tag{5-12}$$

provided that z is sufficiently small compared to the range over which Z varies, so that $z^2 d^2g/dZ^2$ is negligible.

Since dg/dZ is negative, $(dv_z/dt)_d$ is negative for positive z and positive for negative z; i.e., the acceleration tends to compress the cluster in the z direction. It is the gravitational force produced by disc material between $Z_C + z$ and $Z_C - z$ that is responsible for this inwards acceleration of cluster stars at $\pm z$. This acceleration increases the average stellar energy, providing an important heat source for the cluster.

Since the energy change of the cluster occurs in a brief transient pulse, this heating process is called a shock. Such a shock, caused by gravitational perturbations, has little in common with shock fronts in fluid dynamics other than their short transient character. Shocks in a cluster caused by passage through the galactic disc are called "compressive gravitational shocks," in contrast with the "tidal gravitational shocks" which are produced by encounters of a cluster with other spherical masses—see §5.2c.

The change of velocity, Δv_z, experienced by a star as a result of a compressive shock may be computed readily on the impulsive approximation. This approximation assumes that during the perturbation the motion of the star with respect to the cluster center is negligible. On this basis, z is a constant in equation (5-12). In addition we may assume that Z_C is a known function of time; as we shall see below, a cluster approaching the galactic plane will generally have a Z velocity of at least 100 km/s, which we denote by V and, for simplicity, take to be positive; any further changes in V as the system moves through the disc are relatively small. Hence we can replace Z_C by Vt in equation (5-12). With these two assumptions, we can integrate directly, and the resultant velocity change is given by

$$\Delta v_z = \frac{z\,\Delta g}{V} = -\frac{2zg_m}{V}. \tag{5-13}$$

We denote by g_m the maximum value of $|g|$ reached well above or below the galactic disc. (We ignore here the contribution to ϕ from the galactic center, which produces a small but steady increase of $\partial\phi/\partial z$ as z increases.) From equation (5-13) it follows that $(\Delta E)_{\text{Av}}$, the change of energy per unit mass for stars at a particular z, averaged over a symmetrical initial distribution of v_z values, is given by

$$(\Delta E)_{\text{Av}} = \frac{2z^2g_m^2}{V^2}. \tag{5-14}$$

For any particular star the energy change resulting from the perturbation will usually be dominated by the $v_z \Delta v_z$ term; since the mean v_z is zero, this term vanishes on the average and does not appear in equation (5-14).

Before we apply equation (5-14) to compute the heating of actual clusters, we investigate the range of validity of the impulsive approximation. For this purpose we idealize the problem so that $(\Delta E)_{\text{Av}}$ may be computed exactly and compared with equation (5-14). The primary simplifying assumption is that the gravitational potential of the cluster, which we denote by $\phi_C(r)$, is parabolic; i.e., $\phi_C(r)$ is proportional to r^2. In this potential all stars, in the absence of perturbations, oscillate back and forth in simple harmonic motion with the same angular frequency ω. Motions along the three axes are completely uncoupled. We adopt equation (5-12) for the perturbing acceleration in the z direction. The general equation of motion for a cluster star may then be written, if we substitute from equation (5-11) for dg/dZ in equation (5-12),

$$\frac{d^2z}{dt^2} + \omega^2 z = -4\pi G\rho_d(Z)z = \lambda\omega z F(t). \tag{5-15}$$

In this equation λ is a small dimensionless parameter, proportional to g_m, and $F(t)$ is a known function, normalized so that its integral over all time is unity.

This equation may be solved by successive approximations to any order in λ. We consider terms through λ^2, since the energy gain is proportional to λ^2, and write

$$z = z_0 + \lambda z_1 + \lambda^2 z_2 \ldots, \tag{5-16}$$

where

$$z_n = A_n \cos\omega t + B_n \sin\omega t. \tag{5-17}$$

In equation (5-17) A_0 and B_0 are constants, while for higher orders A_n and B_n are functions of time. Combining terms of order λ^n in equation (5-15) and solving the differential equation for each z_n by the method of variation of parameters, we obtain the recursion formulae

$$\begin{aligned} A_n(t) &= -\int_{-\infty}^{t} F(\tau)\sin\omega\tau\, z_{n-1}(\tau)\,d\tau, \\ B_n(t) &= \int_{-\infty}^{t} F(\tau)\cos\omega\tau\, z_{n-1}(\tau)\,d\tau. \end{aligned} \tag{5-18}$$

To determine the total energy E_z associated with z motion we compute $(dz/dt)^2/2$ for the condition that $z = 0$ through terms of order λ^2. Vanishing z makes it unnecessary to consider specifically the contribution of the potential energy to the total energy. If dz_n/dt is computed from equations (5-17) and (5-18), we find that the time derivatives of A_n and B_n drop out, since $(\cos \omega t)\, dA_n/dt$ equals $-(\sin \omega t)\, dB_n/dt$. The expression for $(dz/dt)^2$ simplifies when we use the vanishing of z to determine $\tan \omega t$ and hence the other trigonometric functions of ωt. After some algebra we obtain

$$\left(\frac{dz}{dt}\right)^2 = \omega^2\left[\left(\sum_{n=0}^{2} \lambda^n A_n\right)^2 + \left(\sum_{n=0}^{2} \lambda^n B_n\right)^2\right]. \tag{5-19}$$

For large t, when $F(t) = 0$ and the perturbation has ended, A_n and B_n are constant. The energy change ΔE_z is then also constant and may be computed from equation (5-19) by subtracting $\omega^2(A_0^2 + B_0^2)$ and dividing by 2. For all stars with some particular $\langle z^2 \rangle$ we can now average ΔE_z over all relative phases of the z oscillation. For random initial phases we have

$$\langle A_0^2 \rangle = \langle B_0^2 \rangle = \langle z^2 \rangle, \tag{5-20}$$

and also

$$\langle A_0 B_0 \rangle = 0. \tag{5-21}$$

If we express A_1 and B_1 in terms of A_0 and B_0, using equation (5-18), we find that because of equations (5-20) and (5-21) the A_0A_1 term in equation (5-19) cancels that in B_0B_1 and we are left with

$$(\Delta E_z)_{\text{Av}} = \tfrac{1}{2}\omega^2\lambda^2\langle A_1^2 + 2A_0A_2 + B_1^2 + 2B_0B_2 \rangle. \tag{5-22}$$

The quantities A_1, A_2, B_1 and B_2 are all evaluated at $t = \infty$, and the average is taken over A_0 and B_0, subject to equation (5-20). Equation (5-22) corresponds to the average energy change computed in equation (5-14). Since in any case $\Delta E_x = \Delta E_y = 0$, we replace $(\Delta E_z)_{\text{Av}}$ in equation (5-22) by $(\Delta E)_{\text{Av}}$.

To evaluate equation (5-22) we must assume a specific form for $F(t)$. Again we neglect the change of velocity V as the cluster moves across the disc, and with a Gaussian distribution of density in the galactic disc we obtain

$$F(t) = \frac{V}{\pi^{1/2} H} e^{-V^2 t^2/H^2}, \tag{5-23}$$

while the parameter λ becomes

$$\lambda = -4\pi^{3/2} G\rho_d(0) H/(\omega V) = -2g_m/(\omega V). \tag{5-24}$$

From equations (5-17) and (5-18) it follows that in equation (5-16) the ratio of each term to the preceding one, equal to $\lambda z_n/z_{n-1}$, is less than $|\lambda|$ in magnitude. From equation (5-24) we see that the absolute value of this expansion parameter equals the ratio of the orbital period of a cluster star to π times the acceleration time V/g_m; this ratio is small in most conditions of interest.

The computation of $A_1(t)$ and $B_1(t)$ from equation (5-18), as t approaches ∞, is straightforward. Evaluation of the double integrals for A_2 and B_2 can be achieved on integration by parts. The terms in equation (5-22) combine to yield the final result

$$(\Delta E)_{\mathrm{Av}} = \tfrac{1}{2}\omega^2\lambda^2\langle z^2\rangle I_c^2. \tag{5-25}$$

The quantity I_c is the standard definite integral,

$$I_c = \int_{-\infty}^{\infty} F(\tau)\cos 2\omega\tau\, d\tau = e^{-\beta^2/4}, \tag{5-26}$$

where by definition

$$\beta = \frac{2\omega H}{V}. \tag{5-27}$$

In deriving equation (5-25) we have used the even character of $F(\tau)$ to drop a corresponding integral I_s with $\sin 2\omega\tau$ replacing $\cos 2\omega\tau$. Combining equations (5-24), (5-25) and (5-26) gives

$$(\Delta E)_{\mathrm{Av}} = \frac{2\langle z^2\rangle g_m^2 e^{-\beta^2/2}}{V^2}. \tag{5-28}$$

Equation (5-28) differs from equation (5-14) in two respects. First the average refers to stars with a mean square amplitude of oscillation within the cluster equal to $\langle z^2\rangle$ rather than to stars all of the same instantaneous value of z. Second, this later equation includes the multiplicative factor $e^{-\beta^2/2}$, which equals 0.61 for $\beta = 1$, but drops to 0.135 for $\beta = 2$ and to 3.4×10^{-4} for $\beta = 4$. In this simple case of a parabolic cluster potential the energy is proportional to the action integral; i.e., to the integral of $v_z\, dz$ over the orbit. Equation (5-28) indicates how much this action integral is changed by the assumed perturbation for different values of β.

We investigate more closely the numerical values in equations (5-14) and (5-28). As regards V, different spherically symmetric models [9] of globular cluster motions in the Galaxy give values between 95 and 140 km/s for the rms cluster velocity in either coordinate perpendicular to $\mathbf{R}_G$; the preferred model gives 118 km/s. These results all refer to clusters at a galactocentric distance R_G of 9 kpc. The Z acceleration produced by the galactic disc, which is ignored in these models, would be expected to have some influence on V, the Z velocity with which a cluster crosses the galactic plane. While this acceleration would not affect the rms V for all clusters if the velocity distribution were isothermal, the values of V for those clusters in orbits extending far from the plane would be increased above the values obtained in a spherically symmetric model.

To determine this effect on V we consider the value of g, which is needed in any case for the evaluation of equations (5-14) and (5-28). The Z acceleration in the solar neighborhood is reasonably well known [10]. Half the material in the local galactic disc is within some 150 pc from the galactic plane, producing an acceleration of about 3×10^{-9} cm/s^2. Most of the remaining material in the disc is within some 600 pc of the $Z = 0$ plane. Hence well outside the galactic disc the vertical acceleration g_m is 6×10^{-9} cm/s^2, and this acceleration over a Z distance of 3000 pc, for example, will increase V^2 by 1.1×10^4 km^2/s^2. If this is added to an initial dispersion of 118 km/sec (at $Z \approx 3000$ pc) the resultant rms V is 160 km/s, which we shall adopt as a typical value. Clusters in orbits restricted to low Z will cross the galactic plane near the Sun at lower V. On the other hand, clusters crossing the plane nearer the galactic center will be subject to a greater Z acceleration.

We consider now the magnitude of the shock heating effects to be expected for the values of V and g_m given above and for a cluster with average properties, with M_C equal to $10^5\ M_\odot$, while r_c and r_t equal 1 and 30 pc, respectively; a value of r_h equal to $0.15\ r_t$ or 5 pc is consistent with the King models described in §1.2. To compute β, we may let $H = 300$ pc, corresponding to half the mass within 140 pc for the Gaussian density distribution assumed in equation (5-23). For a star in a circular orbit at the radius r_h, $\omega = 4.3 \times 10^{-14}$/s and for $V = 160$ km/s, $\beta = 5.0$. The quantity $|\lambda| = 2g_m/(\omega V)$ equals 1.7×10^{-2}, indicating convergence of the expansion in equation (5-16).

According to equation (5-28), $(\Delta E)_{\text{Av}}$ for this value of β is reduced by nearly six orders of magnitude below the value in equation (5-14), and for the inner half of the cluster mass, compressive gravitational shocks produce no direct heating. Since β^2 varies about as $1/r^3$, $\beta^2/2$ will be less than 1 for the cluster mass beyond some $2.4r_h$, and equation (5-14), obtained from the impulsive approximation, is applicable only in this outer region. However, this restriction may have only a moderate effect on ΔE_T, the change in total cluster energy. On the impulsive approximation this quantity is given by

$$\Delta E_T = \frac{1}{3} M_C r_m^2 \frac{(\Delta E)_{\text{Av}}}{\langle z^2 \rangle} = \frac{2}{3} M_C r_m^2 \frac{g_m^2}{V^2}, \tag{5-29}$$

where M_C is the total cluster mass and r_m^2 is the mean square radius for the cluster as a whole, equal to three times the corresponding average of z^2. Since it is the regions well outside r_h which usually make the preponderant contribution to r_m^2, the impulsive contribution may give a reasonable approximation for the overall shock heating rate in many clusters.

The velocity change Δv_z produced by this effect is not very large, amounting to only 0.3 km/s for a star with $z = 15$ pc, with $|g_m|$ and V equal to 6×10^{-9} cm/s^2 and 160 km/s, respectively. However, these changes occur twice during each orbital period, P_C, of the cluster with respect to the Galaxy.

The actual rate of compressive shock heating in a cluster must be determined from detailed models, such as those described in §5.2b below. To provide a convenient and simple parameter in terms of which the model results can be expressed we introduce the "shock heating time," defined by

$$t_{sh} \equiv \frac{-\langle E \rangle}{dE_h/dt}. \tag{5-30}$$

In this expression $\langle E \rangle$ is the mean energy per unit mass of the cluster stars and dE_h/dt is the mean rate of shock heating for stars at the half-mass distance r_h from the cluster center; this rate is computed on the impulsive approximation. Since dE_h/dt equals $(\Delta E)_{\mathrm{Av}}$ in one cluster passage through the disc, divided by $P_C/2$, equation (5-14) yields

$$t_{sh} = \frac{3GM_C P_C V^2}{20 g_m^2 r_h^3}. \tag{5-31}$$

As before we have assumed that the mean value of z^2 at $r = r_h$ equals $r_h^2/3$, and have eliminated $\langle E \rangle = (T + W)/M$ with use of equation (1-10).

b. Compressive shocks and cluster evolution

Two sets of Princeton Monte Carlo models give information on the evolutionary effects produced by compressive gravitational shocks. In one of these [5] all stars were assumed to have the same mass, while in the other [7], three mass components were assumed. A tidal cut-off radius, r_t, was assumed in all the models. Equation (5-14), based on the impulsive approximation, was used to compute the velocity perturbation which was applied at intervals to each shell of stars. To fit heating in the z direction into a spherically symmetric model, the value of Δv_z given by this equation was resolved into components, Δv_r and Δv_θ, proportional to $\cos \theta$ and $\sin \theta$, respectively, with $\cos \theta$ chosen at random. Since the time interval between such perturbations was generally less than the orbital period, successive values of Δv_r were taken to be alternatively positive and negative, to avert a change of energy proportional to $\sum v_r \Delta v_r$, summed over successive compressive shocks. In an actual cluster, the successive perturbations to v_z are all negative, but are separated in time by many orbital periods, with the result that $v_z \Delta v_z$ is as often positive as negative, and the energy growth is proportional to $0.5 \sum (\Delta v_z)^2$ rather than to $\sum v_z \Delta v_z$.

A primary result given by these models is the effect of compressive gravitational shocks on the rate at which stars escape from the cluster. This process is presumably caused primarily by the random walk in energy, which is correctly given in these model computations, but is ignored in equation (5-14). For systems with a single mass component, models were obtained [5] with r_h/r_t equal to 0.11 and 0.34 and with values of t_{sh}/t_{rh} ranging from 0 to ∞. The observed values of dN/dt, the rate at which stars escaped from each model, were fitted (within 25 percent) by the equation

$$-\frac{1}{N}\frac{dN}{dt} = \frac{\alpha}{t_{rh}} + \frac{\gamma}{t_{sh}}. \tag{5-32}$$

As noted already in §5.1b, α increases from 0.015 to 0.05 as r_h/r_t increases from 0.11 to 0.34; for infinite r_t, α is about 4×10^{-3} (see §4.2b). When a small correction is made for the approximate nature of the models, in which the relative change of energy of the halo stars per orbit was not small, the value of γ is independent of r_h/r_t and equals 2.

When stars of different masses are present, the presence of shock heating increases the loss rate for all components. Models with different r_h/r_t and different mass spectra give somewhat different results [7], but in all cases the loss rate for the heaviest stars decreases with time as these stars become concentrated in the inner regions.

A second interesting set of results obtained from these models is the effect of compressive shock heating on core collapse. For the one-component models with $r_h/r_t = 0.34$, a value of t_{sh} less than $3.8t_{rh}$ leads to an expansion of all regions of the cluster, including even the core. This result would likely be modified if deviations from the impulsive approximation were taken into account. In an actual cluster the zero-order orbital frequency increases rapidly as r decreases, with a corresponding increase in β according to equation (5-27). Hence the orbits of stars in the inner regions will change nearly adiabatically and the direct heating rate for such stars will be very low.

For t_{sh} exceeding $3.8t_{rh}$ the cluster cores collapse as expected. In the one-component models with $r_h/r_t = 0.11$, the value of t_{coll}, the time to collapse, expressed in units of t_{rh}, decreases from 15 to 9 as t_{sh}/t_{rh} decreases from ∞ to 17. Within this range of parameters, shock heating actually accelerates core collapse. This effect may be understood at least in part as a result of the depletion in halo population. Shock heating accelerates loss of stars by the halo and consequently steepens the gradient of the velocity distribution function. As a result the rate at which stars diffuse out of the core by dynamical relaxation is increased, with a consequent rise in the core contraction rate at early evolutionary phases. A similar effect was noted in §4.2a; Model F, with no initial halo, evolved more rapidly than Model D, which had an extensive halo resulting from an early dynamical collapse. As a result $t_{\text{coll}}/t_{rh} = 12$ for Model F as compared to 18 for Model D.

This variation of t_{coll}/t_{rh} with t_{sh}/t_{rh} is not present in the multi-component models, where the development of mass stratification, rather than evaporation from the halo, is presumably the main cause of early core contraction. As in the isolated clusters discussed in §4.2d the presence of several mass components reduces t_{coll}/t_{rh} to values mostly between 2 and 4.

The later stages of core collapse in these various tidally truncated models may be attributed to the gravothermal instability, although this explanation lacks the relatively firm foundation obtained from the detailed models of isolated clusters—see §§4.2c and 4.2d.

An approximate computation of t_{sh} for 32 globular clusters [11] gives values which range between 10^{10} and 10^{14} years; only in 3 of the 32 is t_{sh}/β smaller than t_{rh}/α. Hence for most clusters at the present time shock heating does not greatly affect the rate of stellar escape, though it may speed on their way stars in the outer halo which are diffusing up to the escaping energy. A more precise theory would likely give a greater spread in true shock heating times, with deviations from the impulsive approximation leading to some increases, while for clusters crossing the galactic plane at smaller R_G a greater g_m may decrease t_{sh}. If some clusters were present initially with relatively low mean densities, such rarefied systems may have dissipated entirely. Thus it is possible [11,12] that stars now moving freely in the galactic halo were mostly formed in clusters which were then disrupted. Such a cluster origin is relatively more likely for stars of lower mass, whose escape has left a preponderance of somewhat heavier stars remaining in the clusters. More detailed models are required to show quantitatively the conditions necessary for such processes.

c. Tidal shocks

When a cluster is perturbed by the tidal gravitational force of a passing spherical mass, the process may be termed a "tidal gravitational shock," often abbreviated to "tidal shock." As with any transient gravitational field produced by an external source, the cluster stars will gain energy on the average. Evidently tidal shocks, like compressive shocks, can heat a globular cluster. Random walk of energy resulting from such shocks also enhances the rate of stellar escape.

As in §5.2a, the change of stellar energy can be computed simply on the impulsive approximation, neglecting the change of stellar positions in the cluster during the passage of the perturbing mass, whose velocity relative to the cluster center we denote by $\mathbf{V}$. To simplify the analysis further, we assume also that $\mathbf{V}$ is constant both in direction and in magnitude. The distance of closest approach in this rectilinear path we again denote by p.

The stars will be assumed relatively motionless in an inertial reference frame and hence we integrate the equations of motion in such a frame. However, the tidal force, $\mathbf{F}_t$, is written most simply in the rotating frame introduced in §5.1, which we denote here by x', y' and z'; the x' axis points toward the perturber. We ignore the centrifugal term in equation (5-1); recalling equation (1-32) we have the usual results,

$$F'_{tx} = \frac{2x'GM_p}{R^3}, \qquad F'_{ty} = -\frac{y'GM_p}{R^3}, \qquad F'_{tz} = -\frac{z'GM_p}{R^3}. \tag{5-33}$$

Here M_p is the mass of the spherical perturber, whose center is at a distance R from the cluster center. The change of R with time is given by

$$R^2 = p^2 + V^2t^2. \tag{5-34}$$

We introduce now an inertial frame x, y, z, where $\mathbf{x}$, $\mathbf{y}$ and $\mathbf{z}$ are parallel to $\mathbf{p}$, $\mathbf{V}$ and $\mathbf{z}'$, respectively. If we transform equation (5-33) to give $d\mathbf{v}/dt$ in these inertial coordinates, and integrate over t from $-\infty$ to $+\infty$, we find [13]

$$\frac{\Delta v_x}{x} = -\frac{\Delta v_z}{z} = \frac{2GM_p}{p^2V}. \tag{5-35}$$

The component of $\mathbf{\Delta v}$ in the y direction vanishes because of symmetry. As in §5.2a we denote by $(\Delta E)_{\text{Av}}$ the average change of energy for stars with particular values of x and z. This energy equals the corresponding average initial gain in kinetic energy per unit mass, giving

$$(\Delta E)_{\text{Av}} = \frac{x^2 + z^2}{2}\left(\frac{2GM_p}{p^2V}\right)^2, \tag{5-36}$$

The total energy change for the cluster may be obtained as in equation (5-29), substituting $r_m^2/3$ for $(x^2 + z^2)/2$ in equation (5-36) and multiplying by M_C. Thus we obtain

$$\Delta E_T = \frac{M_C r_m^2}{3}\left(\frac{2GM_p}{p^2V}\right)^2. \tag{5-37}$$

As in §5.2a for compressive shocks, an approximate solution may be obtained for slow encounters without use of the impulsive approximation. If, as before, the cluster potential is assumed parabolic, so that all stars oscillate with the same angular frequency ω, this solution assumes a simple analytic form [13]. Comparison of this result with equation (5-36) shows that the impulsive approximation is again valid for ω less than a critical value, equal here to V/p (cf., equation (5-27)); for higher ω, ΔE_T falls rapidly, decreasing by an order of magnitude as ω increases from V/p to twice this critical value.

Equation (5-36) may be used to give approximate results when a cluster in an eccentric orbit passes by the galactic nucleus, approximated here as a point mass. Representing cluster passage through perigalacticon by rectilinear motion, with $p = r_p = a(1 - e)$, is a rather crude approximation but within the framework of the impulsive approximation should give a correct order of magnitude for $(\Delta E)_{\text{Av}}$. We set V in equation (5-36) equal to the pericenter velocity, V_p, which from the usual energy equation—or by combining equations (5-8) and (5-9)—is given by

$$V_p^2 = \frac{GM_G}{a}\left(\frac{1+e}{1-e}\right). \tag{5-38}$$

As before we modify equation (5-36) to give the mean increase of energy per unit mass for stars in an orbit with an rms distance r from the cluster center and find

$$(\Delta E)_{\text{Av}} = \frac{4}{3}\frac{GM_G r^2}{a^3(1-e)^3(1+e)}. \tag{5-39}$$

This equation, which is inapplicable if e is appreciably less than 0.5, certainly much exaggerates ΔE in those situations where the impulsive approximation is invalid.

More precise values of $(\Delta E)_{\text{Av}}$ have been obtained by integrating the dynamical equations for individual cluster stars as the cluster orbits about a central point mass, with $e = 2/3$. These simulations [3], which gave data on stellar escape, referred to in §5.1a, also yielded results on the changes of energy experienced by different stars. The value of x_e given by equation (5-10) equals 53 pc at pericenter for the parameters assumed. According to equation (5-39), $(\Delta E)_{\text{Av}}$ for r equal to this distance is 11 (km/s)^2 corresponding to $\langle(\Delta v)^2\rangle^{1/2}$ about half the circular velocity of 8 km/s at $r =$ 53 pc. The energy changes obtained in the simulations are of this general order, but within each group of stars with a particular orbit the individual values of ΔE are very different for different stars, resulting presumably from differences of orbital position during the one perigalacticon passage considered. The $\mathbf{v} \cdot \mathbf{\Delta v}$ term and the $\frac{1}{2}(\Delta v)^2$ term are of comparable magnitude in these simulations, and the 15 stars considered in each orbit provide too small a sample to average out the first term and obtain a reliable value of the second one.

A noteworthy feature of the calculations is the large difference between direct and retrograde orbits, already noted in the discussion of stellar escape—see §5.1a. Direct circular orbits show a large range of ΔE, from

-20 to more than $+30\,(\text{km/s})^2$, while for the retrograde circular orbits $|\Delta E|$ is generally less than $5\,(\text{km/s})^2$. For stellar orbits with $e = 0.8$, however, the differences between direct and retrograde orbits are small and $\langle(\Delta E)^2\rangle^{1/2}$ is about $10\,(\text{km/s})^2$, again for $r_a = 50$ pc.

It is evident both from theory and from simulation that for a cluster in an eccentric orbit, passages through perigalacticon can have a significant heating effect on the stars in the outer halo of a globular cluster. However, such tidal shocks, like the compressive shocks discussed above, are not likely to have much effect on most of the cluster mass. At the radius $r_t = x_e$ equation (5-10) shows that M_G/R_p^3 is comparable with M_C/r_t^3. It follows that the angular velocity of a star in a circular orbit at the radius r_t is comparable with V_p/R_p. For circular orbits at a smaller radius r, the angular velocity increases as $1/r^{3/2}$, and becomes appreciably larger than V_p/R_p. The impulsive approximation is then a gross overestimate and the heating rate becomes quite small. Heating by passage through perigalacticon is likely to be restricted to the outer half of the cluster's radius, containing only a fraction of the mass—three percent in the Hénon tidally truncated model, for example.

Heating by tidal shocks also occurs when a globular cluster passes by other clusters or by gas clouds in the galactic disc, especially the giant molecular clouds. Equation (5-37) may be multiplied by $2\pi n V p\, dp$ and integrated over dp to find dE_T/dt, the change of energy per unit time resulting from perturbers of volume density n. Encounters with an impact parameter p less than the sum of the radii of the two interacting systems must be considered separately, since equation (5-37) becomes invalid; these give a contribution roughly equal to that of the more distant encounters. The resultant formula [5] for the system disruption time has been applied to galactic clusters, for which disruption by giant molecular clouds can be important [14]; because of the great mass of one such cloud, disruption in a single impact becomes dominant for the less dense clusters. However, for globular clusters, with their higher internal density and higher velocity relative to the gas clouds, disruption by such clouds appears unimportant. Similarly, the shock heating time t_{sh} for globular clusters by shock heating in encounters with each other exceeds the ages of the clusters by some three to four orders of magnitude, and this process may be ignored.

As a result of these perturbations by gravitational shocks, both compressive and tidal, the velocity distribution of stars in a cluster halo will be modified. In particular the predominance of radial orbits in the halo, a result well established for isolated clusters—see §4.2a—will be somewhat altered. During cluster evolution stars will be supplied to the halo in radial orbits, with a low angular momentum J, but before such stars escape, their J values will increase because of random walk produced by these external perturbations. Since these effects have not been explored in the available models, there are no definite predictions to compare with the observed velocity anisotropy in the outer regions of M3, for example.

REFERENCES

1. F. R. Moulton, *An Introduction to Celestial Mechanics*, 2nd edition, 1914 (Macmillan, N.Y., reprint by Dover, N.Y., 1970), Art. 158.
2. I. R. King, *A. J.*, **67**, 471, 1962.
3. D. W. Keenan and K. A. Innanen, *A. J.*, **80**, 290, 1975.
4. K. A. Innanen, W. E. Harris and R. F. Webbink, *A. J.*, **88**, 338, 1983.
5. L. Spitzer and R. A. Chevalier, *Ap. J.*, **183**, 565, 1973.
6. M. Hénon, *Ann. d'Astroph.*, **24**, 369, 1961.
7. L. Spitzer and J. M. Shull, *Ap. J.*, **201**, 773, 1975.
8. J. S. Stodółkiewicz, *Dynamics of Star Clusters*, IAU Symp. 113, ed. J. Goodman and P. Hut (Reidel, Dordrecht), 1985, p. 361.
9. C. S. Frenk and S.D.M. White, *M. N. Roy. Astron. Soc.*, **193**, 295, 1980.
10. J. N. Bahcall, *Ap. J.*, **276**, 169, 1984.
11. J. P. Ostriker, L. Spitzer and R. A. Chevalier, *Ap. J.* (*Lett.*), **176**, L51, 1972.
12. S. M. Fall and M. J. Rees, *M. N. Roy. Astr. Soc.*, **181**, 37P, 1977.
13. L. Spitzer, *Ap. J.*, **127**, 17, 1958; see also E. Knobloch, *Ap. J.*, **209**, 411, 1976. [In the earlier paper a typographic error interchanged the β and β^2 factors in the first term of equation (36)]
14. R. Wielen, *Dynamics of Star Clusters*, IAU Symp. 113, ed. J. Goodman and P. Hut (Reidel, Dordrecht), 1985, p. 449.

6

Encounters with Binary Stars

Binary star systems can provide an important source of energy for globular clusters. The binding energy of such a system can be increased during an encounter with a single star; this decrease in binary energy is matched by a corresponding increase in the kinetic energy of relative motion of the single star with respect to the binary. Under some conditions, the resultant increase of internal kinetic energy of the cluster, computed with each binary regarded as a mass point, can have important dynamical consequences on the cluster's evolution. Some binaries may be primordial; i.e., they may have formed at the same time as most of the cluster stars. A minimum population of binaries is that produced by ongoing processes within the cluster.

In this chapter we discuss the relevant physical processes which involve binary stars. In particular, §6.1 discusses the cross-sections and corresponding rate coefficients for encounters between a star and a binary, leading to changes in energy and eccentricity. Considerable understanding can be achieved with the use of an approximate theory. However, precise results are obtained only by extensive numerical integrations of these three-body encounters. Section 6.2 discusses the more complex reactions possible when two binaries interact. In this situation the available theory is considerably more limited than for a single star interacting with binaries, and dynamical integrations are again the primary source of reaction rates. The formation of binaries from single stars, either by three-body processes or by tidal capture in a close encounter between two stars, is analyzed in §6.3. The rates of these formation processes can be evaluated theoretically with some accuracy.

6.1 ENCOUNTERS OF A SINGLE STAR WITH A BINARY

Before discussing encounter processes we review a number of basic properties of binary systems. The relative orbit of such a system is defined by equation (2-1), with the eccentricity, e, less than 1. The semi-major axis a of the resultant elliptical orbit equals half the average of the minimum and maximum values of r, giving

$$a = \frac{J^2}{G(m_1 + m_2)(1 - e^2)}, \tag{6-1}$$

where we have replaced the masses m and m_f in equation (2-1) by m_1 and m_2. The total energy is again expressed by equation (2-2), which may be simplified as before with the use of equation (2-1). If we eliminate $1 - e^2$ with the aid of equation (6-1), and multiply E by the reduced mass, $m_1 m_2/(m_1 + m_2)$, to obtain the total energy, which we denote by $-x$, we obtain

$$x = \frac{Gm_1m_2}{2a}, \tag{6-2}$$

the net binding energy of the binary.

Next we discuss the population of binary stars in kinetic equilibrium, when energy is freely exchanged in a system of point masses, and detailed balancing exists between each process and its inverse; i.e., when the rates of these two processes per unit volume are equal. Kinetic equilibrium in such a system corresponds to thermodynamic equilibrium in an atomic system, and for the distribution of binary energies and eccentricities we can use the well known corresponding results for the hydrogen atom in the limit of large quantum numbers. We define $f_b^{(0)}(x,e)\,dx\,de$ as the number of binary systems in kinetic equilibrium per unit volume with binding energy and eccentricity within the intervals dx and de centered at x and e. We have

$$f_b^{(0)}(x,e)\,dx\,de = Cg_{n,l}e^{x/kT}\,dn\,dl, \tag{6-3}$$

where $g_{n,l}$, the weight of each state with quantum numbers n and l, is given by

$$g_{n,l} = 2l + 1, \tag{6-4}$$

and C is a constant. We omit the factor 2 which electron spin introduces into $g_{n,l}$ for electron states.

For large l we may write

$$g_{n,l} \propto J, \tag{6-5}$$

where J is again the angular momentum per unit mass; also, we have the usual energy equation

$$x \propto 1/n^2. \tag{6-6}$$

If we use equations (6-6) and (6-1) to determine dx/dn and $(\partial e/\partial J)_a$, and substitute for a from equation (6-2), equation (6-3) yields

$$f_b^{(0)}(x,e) = Ke^{\beta x}e/x^{5/2}; \tag{6-7}$$

by definition, as in equation (1-22),

$$\beta \equiv \frac{1}{kT} = \frac{3}{m_j v_{mj}^2} = \frac{B_j}{m_j}, \tag{6-8}$$

where v_{mj} is the rms random velocity of component j, with particle mass m_j, and B_j is the value for component j of the quantity B defined in equation (1-22). If equipartition is not present, β_j denotes the value of β for each component. By an extension of these arguments [1], one can obtain

$$K = n_1 n_2 G^3 (\pi\beta)^{3/2} (m_1 m_2)^3; \tag{6-9}$$

n_1 and n_2 are the spatial densities of freely moving single stars of masses m_1 and m_2. If particles 1 and 2 are identical, K must be divided by 2 to avoid counting each binary twice.

Often one is interested in the distribution of x without regard to eccentricity. The appropriate distribution function $f_b^{(0)}(x)$ is obtained by integrating $f_b^{(0)}(x,e)\,de$ over all e, which replaces the factor e in equation (6-7) by 1/2.

It may be noted that the distribution function diverges for x either large or small. As with the hydrogen atom, the divergence for small x can be eliminated by the fact that for sufficiently small x the semi-major axes become very large, and the orbits lose their identity either because of perturbations by other stars or because of tidal forces. The divergence at large positive βx is a real effect, and indicates that statistical equilibrium cannot be attained for tightly bound binaries.

An important property of a binary system is the relationship of its binding energy, x, to the kinetic energy of the single stars surrounding it. If x much exceeds $1/\beta = mv_m^2/3$, where v_m^2 is the mean square stellar velocity, the binary is said to be "hard," while in the opposite limit the binary is "soft." Binaries with βx in the neighborhood of 1 have properties intermediate between hard and soft systems.

A sharper boundary exists if one considers interactions of a binary with a single star, of mass m_3, whose initial velocity relative to the center of gravity of the binary we denote by V. The reduced mass, m_r, for the relative motion of the star with respect to the binary is given by

$$m_r = \frac{m_3(m_1 + m_2)}{m_1 + m_2 + m_3}. \tag{6-10}$$

The kinetic energy of this motion, measured at infinite separation, equals $m_r V^2/2$, and equals the binding energy x when V equals a critical value V_c. We see that

$$V_c^2 = \frac{2(m_1 + m_2 + m_3)x}{m_3(m_1 + m_2)}. \tag{6-11}$$

If V is less than V_c the star cannot disrupt the binary; if V exceeds V_c the binary cannot capture the star.

a. Theory of star-binary interaction

When a star encounters a binary, a variety of different processes are possible. The star can remain free, but with the kinetic energy of its motion relative to the binary, evaluated at infinity, increased by an amount y, which can be positive or negative. The energy of the binary will be correspondingly decreased, with the binding energy increased to $x + y$. Such an encounter is called a fly-by. If the initial kinetic energy exceeds x, with $V > V_c$, $x + y$ can be negative, in which case the binary is dissociated. Another possibility is an exchange process, in which one of the binary stars escapes, while the incoming star replaces it in the binary system. If $V < V_c$, a further possibility, in addition to a fly-by or a prompt exchange, is capture of the single star, forming a three-body system, which will generally disrupt sooner or later, with one of the three stars escaping to infinity. This process of interim capture is called "resonance"; the end result is either a resonant fly-by, if the star which was initially free finally escapes, or a resonant exchange, if one of the other two stars leaves the system.

The rates of these different processes are described by differential cross-sections. We let $d\sigma/dy$, a function of x and y, as well as V, be the differential cross-section per interval dy for an encounter in which the binding energy increases by an amount between y and $y + dy$. The reaction rate function $Q(x,y)\,dy$, for y within the interval dy, is defined by

$$Q(x,y) \equiv \int_0^\infty VF(\mathbf{V}) \frac{d\sigma}{dy}(x,y,V)\,\mathbf{dV}, \tag{6-12}$$

where $F(\mathbf{V})$ is the velocity distribution function (normalized to unity), and $\mathbf{dV}$ equals $4\pi V^2\,dV$. The average change of x with time, not including dissociation of the binary, is given by

$$\left\langle \frac{dx}{dt} \right\rangle = n_s \int_{-x}^\infty yQ(x,y)\,dy, \tag{6-13}$$

where n_s is the number density of single stars. For simplicity we generally omit the brackets from dx/dt.

Since soft binaries have relatively little effect on the energy budget of a cluster, the primary emphasis here will be on rate coefficients for processes involving hard binaries. The cross-sections for these processes all include a factor which allows for gravitational focussing of the single star. To derive this factor approximately, we consider an encounter of a single star, with mass m_3 and initial velocity V relative to a spherical mass M. We denote by $r_{\min}$ the minimum distance between the two masses during the encounter and by $V(r)$ the relative velocity when the separation equals r. If p_1 is the impact parameter of the relative orbit whose pericenter distance is $r_{\min}$, we have from conservation of angular momentum and energy

$$p_1 V = r_{\min} V(r_{\min}),$$
$$\frac{1}{2} m_r V^2 = \frac{1}{2} m_r V(r_{\min})^2 - \frac{G m_3 M}{r_{\min}}, \tag{6-14}$$

where we have again introduced the reduced mass, m_r. Eliminating $V(r_{\min})$ and substituting for m_r from equation (6-10), with M replacing $m_1 + m_2$, gives for the encounter cross section σ

$$\sigma = \pi p_1^2 = \pi r_{\min}^2 \left[1 + \frac{2G(M + m_3)}{r_{\min} V^2} \right]. \tag{6-15}$$

This equation is valid for two spherical masses. A binary star, with a semi-major axis a, will appear as a point mass at distances large compared to a; hence for $r_{\min}$ appreciably exceeding a, equation (6-15) gives the cross-section for a star to pass within a distance $r_{\min}$ from the binary's center of gravity.

To obtain the cross-section for some physical process, equation (6-15) must be multiplied by the probability of that process. Unfortunately, such probabilities cannot be computed analytically with any precision. Instead we use detailed balancing in kinetic equilibrium, together with an assumed functional form for the stellar capture cross-section, to determine the dependence of an interaction rate on x and y [1]. The constant in this relationship must be determined from numerical integrations.

Specifically, we evaluate the form of the rate function $Q(x,y)$ for the process of resonance, in which a star is captured by a binary with initial binding energy x. This process is followed in time by the disruption of the triple system, leaving a binary with final binding energy $x + y$. We restrict ourselves to the limit in which the binary is very hard initially as well as finally. We assume for simplicity that the three stars are identical, each with mass m.

The distribution of y values in a resonant process is determined by the way in which the three-body system disrupts. We denote by z the binding energy of the triple star, by w the kinetic energy of relative motion between the binary and the escaping star at infinite separation, and by $u = z + w$ the binding energy of the remaining binary. We define $h(z,u)$ as the normalized distribution function for u with a given z; i.e., $h(z,u)\,du$ is the fraction of triple systems with energy z that will disrupt leaving a binary with binding energy between u and $u + du$. If $Q_d(z,u)$ represents the probability per unit time for disruption of triple systems with energy z, leaving binaries of binding energy

u (per increment of u), the normalized distribution function is given by

$$h(z,u) = \frac{Q_d(z,u)}{\int_z^\infty Q_d(z,u)\,du}. \tag{6-16}$$

Since the relative motions of the reaction products carry away positive kinetic energy w, the final binding energy u cannot be less than z.

This discussion has assumed implicitly that $Q_d(z,u)$ depends only on z and u. In fact, the manner in which triple systems disrupt must depend also on the angular momentum J. We assume here that $Q_d(z,u)$ represents an average over a distribution of J values, excluding the larger values resulting from capture of a star in an extended, weakly bound orbit; thus we assume that the orbits of all three stars pass through the central region of the triple system. We assume also that the motions within the triple are sufficiently close to ergodic so that the distribution of the disruption products depends only on the initial energy and angular momentum.

We now introduce the energy distribution functions $f_t^{(0)}(z)$ and $f_b^{(0)}(u)$ for triples and binaries, respectively, in kinetic equilibrium, while $F_k^{(0)}(w)\,dw$ denotes the fraction of star-binary pairs in kinetic equilibrium for which the initial kinetic energy w of relative motion lies within the interval dw centered at w. Detailed balance between capture and disruption, with w again equal to $u - z$, then gives

$$f_t^{(0)}(z)Q_d(z,u) = f_b^{(0)}(u)F_k^{(0)}(w)n_s[V\sigma_c(u,w)]. \tag{6-17}$$

The term on the left-hand side represents the disruption rate of the triples per unit volume and time, while the right hand side gives the rate of the corresponding capture process. The function $f_b^{(0)}(u)$ is given in equation (6-7), integrated over de, while $F_k^{(0)}(w)$, the normalized Maxwellian distribution function per unit dw, equals $4\pi VF^{(0)}(V)/m_r$, where $F^{(0)}(V)$, the normalized velocity distribution function per unit $\mathbf{dV}$, is given in equations (2-10); n_s is the particle density of single stars. The quantity $\sigma_c(u,w)$ is the cross-section for capture of a single star, whose energy of initial relative motion equals w, by a binary of binding energy u. The increments $dz\,du$ and $du\,dw$ on the two sides of this equation cancel out.

To make use of equation (6-17) we must introduce the functional form of the cross section $\sigma_c(u,w)$. We assume that this cross-section is proportional to πa^2, multiplied by two factors. The first of these is the gravitational focussing factor, which from equation (6-15) is proportional to Gm/aV^2 for a hard binary. The second factor allows for the decrease of $\sigma_c(u,w)$ to zero for $w \geqslant u$, when the total energy of star plus binary is zero or positive.

To evaluate this factor we consider alternative outcomes of the encounter. The kinetic energy of the star when it is near the binary will appreciably exceed w. Capture will tend to occur if the interaction decreases this energy by more than w but less than u; a larger decrease is likely to produce disruption of the binary, with the star bound to one of the binary components, producing a prompt exchange. While this argument is not exact, one may expect that for w somewhat less than u, σ_c will vary as $1 - w/u = z/u$, a variation which we assume for the second correction factor. The capture cross-section obtained with these two correction factors is in rough agreement with numerical results [2] on resonance scattering of stars by soft binaries (i.e., for $w \geqslant u$).

Adopting this form for $\sigma_c(u,w)$ and substituting from equation (6-2) for a in terms of the binary binding energy (designated here by u instead of x), we obtain

$$V\sigma_c(u, w) \propto \frac{m^3 G^2 z}{V u^2}. \tag{6-18}$$

Combining factors on the right-hand side of equation (6-17), we have

$$f_t^{(0)}(z) Q_d(z,u) \propto \frac{z e^{\beta z}}{u^{9/2}}. \tag{6-19}$$

Equation (6-16) now gives

$$h(z,u) = \frac{7}{2} \frac{z^{7/2}}{u^{9/2}}. \tag{6-20}$$

To evaluate $Q(x,y)$ we must multiply $h(z,u)$ by the rate coefficient (6-18) for the formation of triples with binding energy z from binaries of binding energy x (instead of u). Again we denote by V the initial relative velocity of approach between the single star and the binary. The binding energy z of the triple formed is then $z = x - \frac{1}{2}m_r V^2$. When the triple disrupts, the binding energy u of the remaining binary is defined as $x + y$, in accordance with our usual notation.

Finally we must average the resultant expression for $V\sigma_c h(z,u)$ over a Maxwellian distribution for V, keeping x and y constant. This integration is much simplified by the fact that the initial kinetic energy is now very much less than x or z, since the binary is assumed to be hard; this situation differs from the capture considered above in the discussion of detailed balancing, where values of w comparable with but still less than z must be considered. Since $\frac{1}{2}m_r V_1^2$ is relatively small, $z \approx x$; hence we replace z by x in $h(z,u)$, which becomes $h(x,x + y)$ and is independent of V. Similarly, in equation (6-18) for $V\sigma_c$, we replace z by x, with u also replaced by x as noted above. Thus in the integration only the factor $1/V$ changes, and averaging over a Maxwellian distribution gives the following rate function for the resonant process:

$$Q_{\text{res}}(x,y) = AG^2m^3(m\beta_r)^{1/2}\frac{x^{5/2}}{(x+y)^{9/2}}. \tag{6-21}$$

This result assumes that $m_1 = m_2 = m_3$; m_r has been eliminated with the use of equation (6-10). The factor G^2m^3 gives $Q_{\text{res}}\,dy$ the dimensions cm^3/s; the constant A is dimensionless. The quantity β_r is defined in equation (6-8), with $m_rV_m^2$ replacing $m_jv_{jm}^2$. Since V^2 on the average equals $\langle v_s^2 + v_b^2\rangle$, we have

$$\frac{1}{m_r\beta_r} = \frac{1}{m_3\beta_s} + \frac{1}{(m_1+m_2)\beta_b}. \tag{6-22}$$

Here subscripts s and b refer to single stars and binaries, respectively. The quantity β appearing in equation (6-19) may be set equal to β_r. If equipartition is established, as in kinetic equilibrium, $\beta_r = \beta_s = \beta_b$.

The differential cross-section for resonance, corresponding to equation (6-21), may be written in the dimensionless form

$$\frac{1}{\pi a^2}\frac{d\sigma}{d(y/x)} = \frac{2AV_c^2}{(3\pi)^{1/2}V^2}(1 + y/x)^{-9/2}, \tag{6-23}$$

as may be verified by substituting this expression into equation (6-12) and making use of equations (6-2), (6-11) and (2-10), with $j^2 = m\beta_r/3$ in this equal-mass case.

If the motions within the triple system are truly ergodic, the escape probability should be the same for each of the three stars. Thus a third of the resonant events should be classified as fly-bys, with the remaining two-thirds as exchanges. The total reaction rate function for these resonant processes is given by

$$Q_{\text{res}}(x) \equiv \int_0^\infty Q_{\text{res}}(x,y)\,dy = \frac{2}{7x}AG^2m^{7/2}\beta_r^{1/2}. \tag{6-24}$$

The mean value of y is readily found to be

$$\langle y\rangle = 0.4x. \tag{6-25}$$

Equation (6-25) applies only to the close encounters for which equation (6-23) is applicable. When distant encounters are included, the total cross-section, including small values of y/x, is very large, and $\langle y\rangle$ averaged over this cross section is correspondingly small. Negative values of y have been ignored in equation (6-24); the binary energy $-x$ cannot be increased by more than the initial kinetic energy $m_rV^2/2$, and hence y/x cannot be less than $-V^2/V_c^2$, whose absolute value is small for hard binaries.

While the distribution of y values resulting from resonances in close encounters is readily computed from equation (6-23), the final kinetic energies of escaping single star and binary in the reference frame of the cluster depend also on other variables. In the reference frame of the center of mass, star plus binary, the momenta of the star and the binary are equal and opposite, giving kinetic energies inversely proportional to the masses of the two reaction products. In this frame, the final kinetic energies of the binary and the single star are 1/3 and 2/3, respectively, of $y + m_r V^2/2$, if all stellar masses are identical. If one knows the direction of this final velocity $\mathbf{V}$ of separation in the center-of-mass frame and also the velocity of the center of mass relative to the cluster, the final vector velocities can be readily transformed to the cluster frame.

Theoretical cross-sections for hard binaries have also been derived for distant encounters [1], in which a hard binary is perturbed by the tidal force of a passing star. The results of this latter theory are used in §6.1b to obtain a semi-empirical fit for the numerical data on reaction rates.

For soft binaries one result of importance is the rate of dissociation, since this quantity is important in the net rate at which hard binaries form—see §6.3 below. The total cross-section for this process may be computed theoretically [1,3], on the assumption that the single star passes near one of the two stars in the binary and that the other star may be ignored in the dynamics of the encounter. With straightforward analysis one obtains for the dissociation cross-section σ_{dss},

$$\frac{\sigma_{\text{dss}}}{\pi a^2} = \frac{40}{3} \frac{m_3^3}{m_1 m_2 (m_1 + m_2 + m_3)} \frac{V_c^2}{V^2}. \tag{6-26}$$

The dissociation rate coefficient $Q_{\text{dss}}(x)$ is identical in form to $Q_{\text{res}}(x)$ in equation (6-24) with $A = 23.9$ in the equal-mass case. However, the range of validity is different, with $\beta_r x \ll 1$ for Q_{dss}, and $\beta_r x \gg 1$ for Q_{res}.

b. Numerical results for equal masses

Very extensive dynamical calculations, involving orbits for millions of encounters between stars and binaries, have yielded definite results on the probability of energy transfer to passing single stars when all stars have the same mass. The two basic initial parameters in the calculations are V/V_c, which distinguishes the degree of hardness or softness of the binary, and the eccentricity e. The end result, averaged over the other orbital parameters, is the differential cross-section for an increase y in the binding energy. The calculations identify resonance encounters (defined as those during which the quadratic sum of the distances between the three stars displays more than one minimum) and distinguish exchanges from fly-bys, each of which can be resonant or non-resonant.

In the calculations for each V/V_c and e, values of the other orbital parameters, including binary phase, orientation and the impact parameter p, were varied at random with p^2 distributed uniformly from 0 to $p_{\max}^2$, chosen to include essentially all relevant collisions. The differential cross-section $d\sigma/dy$ for a given y is then equal to $\pi p_{\max}^2$, multiplied by the fraction of collisions, per unit dy, for which the binary binding energy after collision lies within the interval dy centered at $x + y$.

Some resultant differential cross-sections for resonant exchanges are shown [4] in Fig. 6.1, where values of the dimensionless quantity $(V/V_c)^2 \times (d\sigma/d\Delta)$, in units of πa^2, are plotted against $\Delta \equiv y/x$. These data were obtained for $e = 0$. Included in the plot are data for encounters with

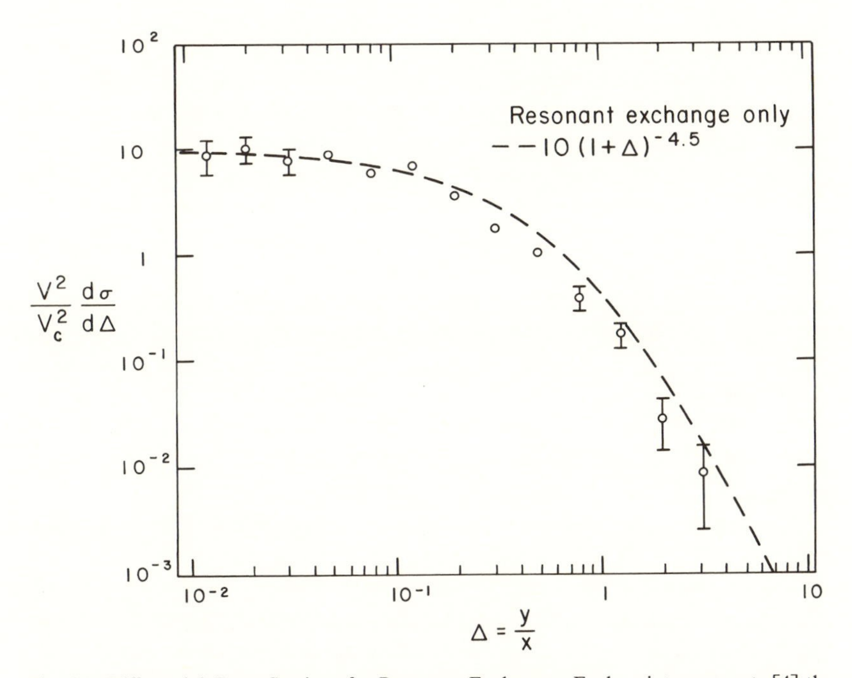

Fig. 6.1. Differential Cross-Sections for Resonant Exchanges. Each point represents [4] the average value of $d\sigma/d\Delta$, where $\Delta \equiv y/x$ is the relative change of the net binding energy of the binary. Each cross-section σ is expressed relative to the geometrical cross-section πa^2, where a is the semi-major axis of the binary. The binaries are all hard, with $V^2/V_c^2 < 0.016$. The initial binary eccentricity e is 0, and all three masses are equal. The error bars (shown when these exceed the diameter of the plotted circles) represent the 1σ statistical deviation expected from the finite number of collisions computed.

$(V/V_c)^2$ in the range between 4×10^{-3} and 0.016. The small fraction of encounters giving y/x negative (but necessarily greater than $-V^2/V_c^2$) are not included. The dashed line represents the theoretical distribution corresponding to equation (6-23). If $d\sigma/d\Delta$ for resonant fly-bys is assumed to be half of that for resonant exchanges, as predicted for close encounters, the value of A in equation (6-21) representing all resonant encounters and corresponding to the dashed line is 23; the best fit for large Δ corresponds to about $A = 12$. In view of the approximate capture cross-section assumed in the derivation, the shape of the theoretical distribution is in remarkable agreement with the data. For higher values of e the agreement is not quite so close.

While the ratio of resonant fly-bys to resonant exchanges is close to the theoretical value of 0.5 for $\Delta > 1$, this ratio rises for small Δ, exceeding 2 for $\Delta < 0.01$. The resonant encounters producing such small changes are primarily distant ones [1]; the captured star is in an extended orbit, well outside the inner binary. This same outer star subsequently escapes with a relatively small gain or loss of kinetic energy from its initial value $m_r V^2/2$. Distant encounters, leading to the capture of such weakly bound stars with relatively large angular momentum, were excluded from the analysis of resonant encounters leading to equation (6-21).

The non-resonant fly-bys tend to have somewhat higher cross-sections than the resonant ones, especially at lower values of Δ, resulting largely from distant encounters. The total differential cross-section, summed over all four processes, resonant and non-resonant fly-bys and exchanges, is shown [4] in Fig. 6.2 for the same range of binary hardness, measured by V/V_c as in Fig. 6.1; the values have here been averaged over $e\,de$—see equation (6-7). Again the few encounters yielding negative Δ (with $|\Delta| <$ 0.016, the maximum value of V^2/V_c^2) are omitted. The plots for other values of V/V_c are almost identical to this figure, except that for $(V/V_c)^2$ in the range from 0.063 to 0.25 slightly more than half the energy changes of the passing star are negative for $|\Delta| \leqslant 0.2$.

The approximate theory for distant encounters [1] predicts a variation of $d\sigma/dy$ that is somewhat less steep than $1/y$. To provide an analytical fit to the observed cross-sections which includes empirically the effect of such encounters, equation (6-23) may be multiplied by $(1 + x/y)^{1/2}$. If we again define Δ as y/x, we obtain the semi-empirical dimensionless formula

$$\frac{1}{\pi a^2}\frac{d\sigma}{d\Delta} = \frac{2AV_c^2}{(3\pi)^{1/2}V^2}\,\frac{1}{\Delta^{1/2}(1+\Delta)^4}. \qquad (6\text{-}27)$$

The solid curve in Fig. 6.2 shows this equation, with A set equal to 21 to yield the best fit to the points.

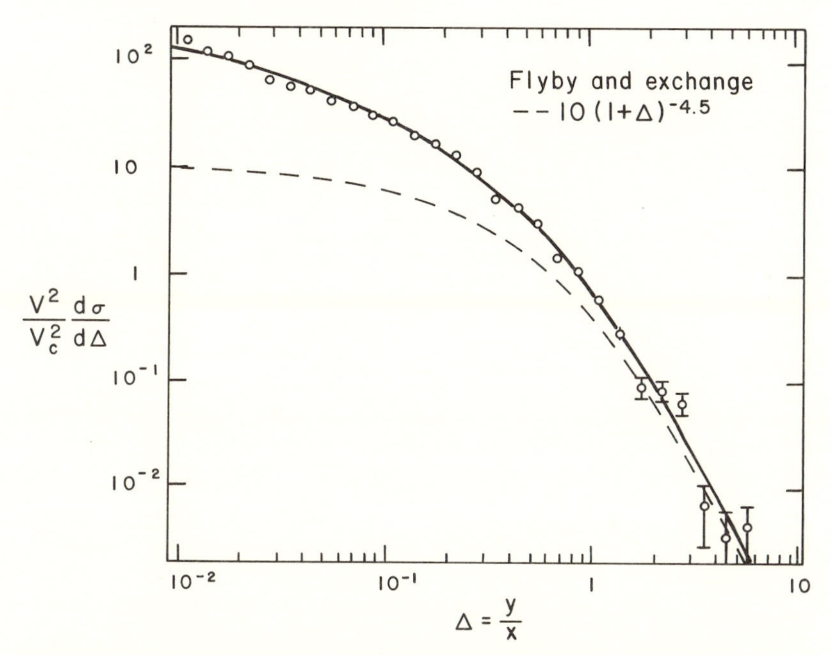

Fig. 6.2. Total Differential Cross-Section. The cross-sections [4], again expressed relative to πa^2, have been summed over resonant and non-resonant collisions and over both fly-bys and exchanges, and have been averaged over a uniform distribution of e^2. Again $V^2/V_c^2 < 0.016$ and all three masses are equal. The solid curve represents equation (6-27), with $A = 21$.

We consider next the energy exchange rate coefficient. For encounters all with the same relative velocity V this coefficient equals $V\langle\sigma y\rangle$, where

$$\frac{\langle\sigma y\rangle}{x} = \langle\sigma\Delta\rangle \equiv \int_{-1}^{\infty} \Delta \frac{d\sigma}{d\Delta}\, d\Delta. \tag{6-28}$$

If this equation is multiplied by $n_s V x$ and averaged over a Maxwellian velocity distribution, one obtains equation (6-13) for the energy loss rate per binary, with $Q(x,y)$ eliminated by use of equation (6-12). For $V > V_c$ the lower limit of integration excludes the dissociation range; for $V < V_c$ the quantity $d\sigma/d\Delta$ must vanish for $\Delta < -V^2/V_c^2$, as noted earlier.

Values of $\langle\sigma\Delta\rangle$ have been computed [5] from data on $d\sigma/d\Delta$ similar to those in Fig. 6.2; the absolute value in units of πa^2 is plotted against V/V_c in Fig. 6.3. As in the previous figure, averages have been taken over a uniform distribution of e^2. Negative values of Δ are included in the integration, of course. For V/V_c less than indicated by the dotted line, at about 0.6, $\langle\sigma\Delta\rangle$ is positive and hard binaries become harder on the average. For V/V_c exceeding this "watershed" value, $\langle\sigma\Delta\rangle$ is negative and soft

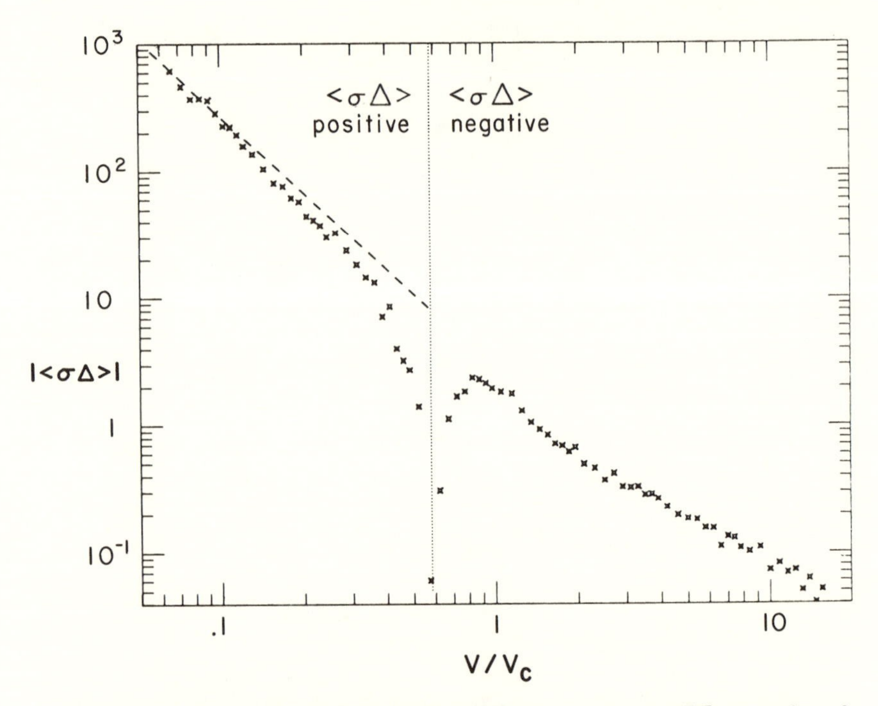

Fig. 6.3. Average Change in Binary Binding Energy. The crosses represent [5] mean values of $(d\sigma/d\Delta) \times \Delta$, integrated over $d\Delta$ and expressed relative to πa^2. The horizontal scale is the dimensionless initial relative velocity, V/V_c. For values of V/V_c to the right of the dotted line, the integral is negative and absolute values are shown. All the processes summed in Fig. 6.2 have been included here. The straight dashed line represents equation (6-29), with $A = 21$.

binaries tend to become softer. Results for individual values of e are very similar to this average curve; the watershed value of V/V_c decreases from 0.67 to 0.54 as e increases from 0 to 1.

For hard binaries, equation (6-27) may be substituted into equation (6-28), with neglect again of the small contribution from negative values of Δ. Integration then yields

$$\frac{\langle \sigma\Delta \rangle}{\pi a^2} = \frac{1}{8}\left(\frac{\pi}{3}\right)^{1/2} \frac{A V_c^2}{V^2}. \tag{6-29}$$

This function of V^2/V_c^2 has been plotted as a dashed line in Fig. 6.3, with A again set equal to 21. As expected, the fit is good at the low V^2/V_c^2 for which A was evaluated. For $V^2/V_c^2 \approx 0.04$ the effective A is about 16. The inclusion of the $(1 + 1/\Delta)^{1/2}$ factor in equation (6-27) has increased $\langle \sigma\Delta \rangle$ by a factor 1.72 over that obtained from equation (6-23); thus small energy changes, produced chiefly by distant encounters, contribute to the overall energy exchange about 72 percent as much as do the close encounters, on which equation (6-23) is based.

To obtain dx/dt for hard binaries, in accordance with equations (6-12) and (6-13), we may integrate equation (6-29) over a Maxwellian velocity distribution for V, with rms value $V_m = (m_r\beta_r/3)^{-1/2}$—see equation (6-8). With use again of equation (6-10) and (6-11), we obtain in the equal-mass case

$$\frac{dx}{dt} = \frac{\pi}{16} A G^2 n_s m^{7/2} \beta_r^{1/2}. \tag{6-30}$$

The rate function for dissociation of a binary in the equal-mass case, as determined from the numerical calculations [2], is given in Fig. 6.4. The

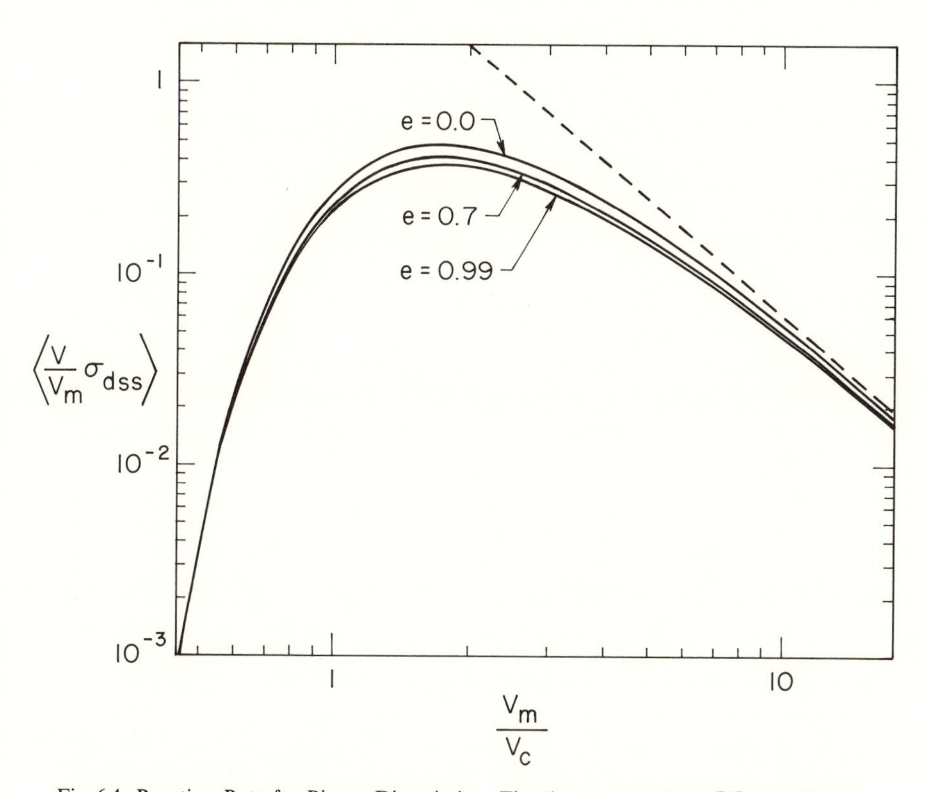

Fig. 6.4. Reaction Rate for Binary Dissociation. The three curves show [2], for different binary eccentricities, the rate coefficient $Q_{dss} = \langle V\sigma_{dss}\rangle$ for the equal-mass case, divided by $V_m \pi a^2$ to provide a dimensionless result. The horizontal scale is V_m/V_c, where V_m is the rms value of the initial relative velocity, assumed to have a Maxwellian distribution. The dashed line represents equation (6-26).

vertical scale represents $Q_{\rm dss} = \langle V\sigma_{\rm dss}\rangle$ in units of $\pi a^2 V_m$. The horizontal scale shows $V_m/V_c = (x\beta_r/1.5)^{-1/2}$. The dashed line shows the theoretical result from equation (6-26), giving $Q_{\rm dss}/(\pi a^2 V_m)$ varying as V_c^2/V_m^2. The asymptotic fit is excellent. The curves for different e are nearly identical.

Many of the numerical investigations indicate that the eccentricity of a binary changes more rapidly than does the binding energy. Quantitative study of this effect, based on a few thousand encounters, shows [6] that as the impact parameter p of an encounter increases, $y \equiv \Delta x$ goes to zero much more rapidly than does the change in e. While more complete numerical results are lacking, it is often assumed that most binaries in globular clusters will approach the equilibrium distribution of eccentricities before their energies are changed much by encounters. Very hard binaries, with a comparable to the stellar radii, behave differently, since tidal forces tend to circularize the orbits.

c. Numerical results for unequal masses

For hard binaries some 5000 orbits of passing stars have been computed [6,7] with a variety of mass ratios. The binary stars were assumed to have equal masses $m_1 = m_2$, and the initial orbits were all circular. In most of the encounters the impact parameter p and the velocity V of the single star were taken to be zero at infinite distance. The values of the energy change, y, were averaged over the phase of the binary and the inclination of its orbit plane relative to the initial direction of the star. In some encounters the effects of finite p and V were explored; however, the small energy changes associated with large impact parameters were not considered.

The results obtained [7] for the mean energy change are plotted in Fig. 6.5 against the mass ratio m_3/m_1. For small m_3/m_1 the energy change varies linearly with m_3. This is to be expected theoretically, since the orbit of a test particle, with $m_3 \ll m_1$, will be entirely independent of m_3, as will its departing velocity at infinity. The value of 0.62 for $\langle\Delta\rangle$ at $m_1 = m_3$ significantly exceeds the value of 0.49 found in a similar calculation [6] for V/V_c between 0.08 and 0.18.

An important result is the dependence of the exchange probability $P_{\rm ex}$ on m_3/m_1. The data in Fig. 6.6 show [7] a very steep increase of $P_{\rm ex}$ as this ratio increases from 0.5 to 1. Other computations also show that in a close encounter with a binary system a heavier star generally tends to displace a lighter star.

The results for $\langle y\rangle$ may be used to compute dx/dt, if the data with finite impact parameter are fitted with the gravitational focussing formula, equation (6-15), in the hard-binary limit. If we assume that $r_{\rm min} = ka$,

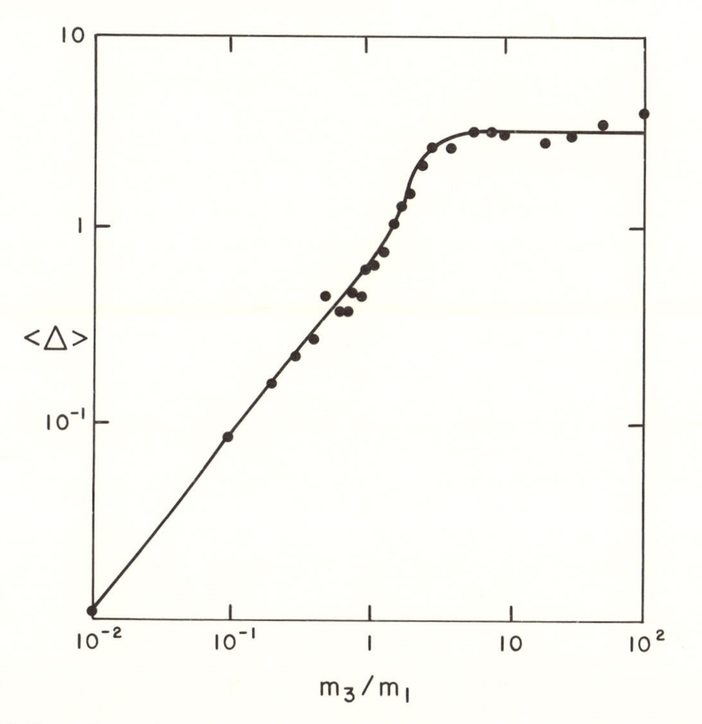

Fig. 6.5. Dependence of Energy Exchange Rate on Stellar Mass. The plotted points show [7] mean values of Δ, the relative increase of binary binding energy, as a function of the mass ratio m_3/m_1, where m_3 is the mass of the incident star and $m_1 = m_2$ is the mass of each star in the binary.

where k is some constant, then with some algebra we obtain

$$\frac{dx}{dt} = \pi k n_s \frac{\langle y \rangle}{x} G^2 m_1 m_2 (m_1 + m_2 + m_3) \langle 1/V \rangle. \tag{6-31}$$

The average of $1/V$ over a Maxwellian distribution equals $(2\beta_r m_r/\pi)^{1/2}$, with β_r given in equation (6-22). The constant k is determined by the condition that equation (6-31), with $\langle y \rangle$ evaluated at $p = 0$, gives the correct result for dx/dt, determined from the data at a variety of p values. The computational results (6) give $k = 0.67$. The corresponding value for dx/dt in the equal-mass case is within 20 percent of that found in §6.1b for resonant encounters alone, but is only about half of that given by equation (6-30) (with $A = 21$), which includes distant encounters.

In the limiting case where $m_3 \ll m_1$ and again $m_2 = m_1$, the discussion above has shown that $\langle y \rangle/x$ is proportional to m_3/m_1; hence we have

$$\frac{dx}{dt} \propto n_s G^2 m_3 m_1^2 \langle 1/V \rangle. \tag{6-32}$$

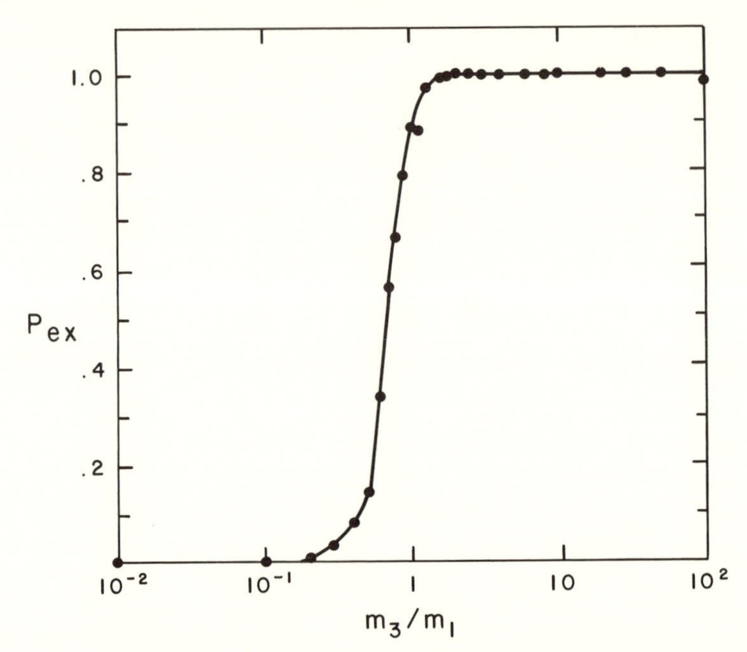

Fig. 6.6. Dependence of Exchange Probability on Stellar Mass. The plotted points show [7] the fraction of collisions with a particular m_3/m_1 which resulted in an exchange, with star 3 remaining in the binary and star 1 or 2 escaping; as before, $m_1 = m_2$.

6.2 ENCOUNTERS BETWEEN BINARIES

When two binaries encounter each other, the range of possible outcomes is greater than for a single star colliding with a binary. We restrict our consideration to hard binaries, which excludes dissociation of both binaries, just as a single hard binary cannot be disrupted by a passing star of average velocity. However, a "disruptive collision" is possible, leading indirectly to dissociation of one of these hard binaries. The two stars released can each depart separately, or one star can be left bound to the remaining binary, with a distance of closest approach sufficiently great so that the resultant three-body system is stable, at least for a substantial period. This disruption process differs from the dissociation discussed above in that the energy source is increased binding energy of the remaining binary rather than initial kinetic energy of translation.

A detailed quantitative theory of these phenomena does not exist. However, an asymptotic theory [8] helps to provide an understanding of the steps involved in a disruptive collision. We assume that of the two binaries, designated as A and B, binary A is much the hardest. All four stars involved are taken to have the same mass, and hence a_A, the semi-major axis of binary A, is much smaller than the corresponding a_B. When the two binaries interact, binary A can usually be regarded as a single star, with a mass twice that of each of the stars in binary B. As shown in Fig. 6.6, the exchange probability is essentially unity for this situation. Hence one of the stars in binary B escapes, leaving the other captured by binary A to form a triple system. If the captured star is in an eccentric orbit, it is likely to experience a close encounter with binary A, in which case it will generally gain enough energy to escape, leaving binary A even harder than before. If the orbit of the captured star has a pericenter distance exceeding a_A by a factor of 3.5 or more, the triple system of equal-mass stars is essentially stable [9] if undisturbed.

To obtain specific rates for these processes, some ten thousand orbits of binary-binary encounters were computed [10]. The masses of all four stars involved were assumed identical, and are denoted here by m. We again use subscripts A and B to denote the harder and the softer binary, respectively. The binding energy ratio Q, defined as $x_A/x_B = a_B/a_A$, was varied systematically from 1 to $2^{3/2}$, with some computations for Q up to $2^{5/2}$. The initial kinetic energy, w (also denoted [10] by T_∞), was varied from x_B to $x_B/16$. This kinetic energy equals $m_r V^2/2$ as before, where V is the relative velocity of the two binaries at infinity; since $m_3 = 2m$ in equation (6-10), $m_r = m$. For each value of Q and w/x_B, several hundred orbits were computed, with e^2, orbital phases and orientation angles varied at random; p^2, the square of the impact parameter, was varied at random from zero to some upper limit, corresponding to a value of $r_{\min}$, the distance of closest approach, roughly equal to $5a_B$. More distant encounters are not often disruptive; fly-bys, which result in two separate binaries, result from more distant encounters as well as from some close ones. While the energy transferred in such binary-binary fly-bys is much the same as in star-binary fly-bys, the average energy released in disruptive encounters between two binaries substantially exceeds that released in a close encounter of a star with a binary. Hence disruptive collisions provide the dominant energy source from binary-binary encounters, and binary-binary fly-bys may be ignored [11].

The cross-section for disruptive collisions, which we denote by σ_{dsr}, can be determined directly from these calculations [10]. If we omit the close encounters that result in fly-bys (13 percent on the average), the data yield the approximate average

$$\sigma_{\text{dsr}} = \frac{16\pi m G a_B}{V^2}. \tag{6-33}$$

If we define V_c^2 as the value of V^2 for which the kinetic energy w equals x_B, and make use of equation (6-11), with m_3 again replaced by $2m$, equation (6-33) can be expressed in the dimensionless form

$$\frac{\sigma_{\text{dsr}}}{\pi a_{\text{B}}^2} = 16\eta \frac{V_c^2}{V^2}; \tag{6-34}$$

the correction factor η has been included to give the observed values of σ_{dsr}, allowing for the small but definite change of $V^2\sigma_{\text{dsr}}$ with the hardness parameter $x_{\text{B}}/w = V_c^2/V^2$ and with the binding energy ratio Q. If η is averaged over Q for each value of x_{B}/w, the values in Table 6.1 are obtained. Since each value in this table is the average of 2000 experiments, an rms accuracy of about two percent is anticipated. If η is averaged over x_{B}/w for each value of Q, the resultant variation of $\langle\eta\rangle$ with Q is small, with a decrease from 1.06 to 0.92 as Q increases from 1 to $2^{3/2}$.

These disruptive collisions lead in some cases to stable triple stars and in others to single remaining binaries. The percentage of triples depends on Q, increasing from about 20 to 50 percent of the total number of collisions (excluding fly-bys) as Q varies from 1 to $2^{5/2}$; the values of x_{B}/w included in these averages are those in Table 6.1. This difference of outcome is probably of little significance for the cluster, since the triple systems are probably disrupted in due course as a result of random walk in eccentricity, resulting mostly from distant gravitational encounters with passing stars—see §6.1b. When the pericenter distance of the captured star is reduced significantly below $3.5a_{\text{A}}$, a close encounter with binary A becomes possible, and one of the three stars escapes with a kinetic energy which can be relatively high. Since detailed rates for eccentricity changes are lacking, this scenario has not been firmly established, but it seems plausible.

The energies released in a disruptive collision are important for cluster dynamics. One would expect that the distribution of kinetic energies for the reaction products would follow equation (6-20), since this result was derived for the disruption of triple systems. Indeed, the data confirm [10] this expectation. We denote by E_4 the energy released in the disruption of the intermediate four-body system, formed at the outset, and by E_3 the energy released in the subsequent disruption of the remaining three-body

TABLE 6.1

Average Correction Factor for Binary-Binary Cross-Section

x_B/w	1	2	4	8	16
$\langle\eta\rangle$	0.79	0.83	0.98	1.08	1.20

system. Then the distribution of E_4 values follows equation (6-20), with $u = z_4 + E_4$, and the same equation applies similarly to E_3, with z_3 replacing z_4. Evidently z_4 and z_3 are the binding energies of the four-body and three-body system, respectively.

Since equation (6-20) is approximate for a three-body system and even more so for a system with four bodies, the effective values of z_3 and z_4 have been determined to give the best fit for the observed distribution of E_3 and E_4, rather than from estimates of the actual binding energies involved. The fitting process gives [10]

$$\frac{1}{z_4} = \frac{1}{1.8}\left(\frac{1}{x_A + w} + \frac{1}{x_B + w}\right), \tag{6-35}$$

$$z_3 = 0.82(E_4 + x_A + x_B - w). \tag{6-36}$$

According to equation (6-25), the mean values of E_4 and E_3 equal 0.4 times z_4 and z_3, respectively.

Equations (6-35) and (6-36) provide a reasonable fit to the data for the range of parameters considered, and their functional form is theoretically plausible. In particular, the quantity in parentheses in equation (6-36) is the actual binding energy of the three-body system. However, the constants in these two equations may be uncertain by as much as 10 percent.

As with the star-binary reactions considered in §6.1, the energies released are divided between the different reaction products in accordance with momentum conservation. Thus after the first disruption of the four-body system, the triple system will have a kinetic energy $E_4/4$ as measured in the reference frame of the center of mass of the two colliding binaries. To obtain the energies in the reference frame of the globular cluster requires assumptions concerning the velocity of this center of mass, both as to its magnitude relative to the cluster and as to its direction relative to the separation velocity of the reaction products. If the triple system disrupts promptly, before its kinetic energy of translation in the cluster has been randomized by gravitational encounters, the division of kinetic energies between binary A and the second escaper will depend also on the angle between the velocities of the two successively escaping stars. These effects, associated with the random velocity of the center of mass, will generally increase the dispersion of kinetic energies gained by the various reaction products.

6.3 FORMATION OF BINARY STARS

To form a binary system from two gravitating masses, with a positive total energy, requires some mechanism to dispose of enough energy so that the two stars will be bound. Two such processes have been considered for conditions in globular clusters. The first is the presence of a third star which can carry away energy in kinetic form. The second is the loss of energy to internal stellar oscillations, excited by tidal forces of each star on the other. These two formation processes, involving three-body encounters and tidal capture, respectively, are considered here.

a. Three-body encounters

The dissociation rate coefficient $Q_{\rm dss}(x)$ shown in Fig. 6.4 may be used with the principle of detailed balancing in kinetic equilibrium to obtain the corresponding rate coefficient $Q_f(x)$ for formation of binaries with binding energy x. Since $Q_{\rm dss}(x)$ varies as $1/x$ for soft binaries—see equation (6-24) and the discussion following equation (6-26)—it is readily shown [1] that $Q_f(x)$ must vary as $x^{-7/2}$, in view of the equilibrium distribution function of binaries given by equation (6-7). For hard binaries Q_f drops even more steeply with increasing x. As a result, very few hard binaries are formed directly by three-body capture.

The dominant process for forming hard binaries by three-body encounters is an indirect one. Soft binaries are formed in large numbers. Most of these are destroyed, but random walk of binding energy in successive encounters will cause some of them to harden sufficiently to cross the watershed energy, and further hardening becomes the rule rather than the exception. Detailed computations [12] show that a binary formed with $\beta x \approx 1$ will have a 10 percent chance of surviving until $\beta x = 10$ by which point the probability of surviving indefinitely (and hardening further at a constant rate) is nearly unity. Most of the binaries that reach $\beta x = 10$ are formed at much lower βx, with some 40 percent formed at $\beta x < 0.1$, even though the survival probability for a binary with this energy initially is only 10^{-4}. The overall net rate of hard binary formation in three-body encounters, which we denote by $(dn_b/dt)_{3\rm b}$, is measured by the steady-state flux of hard binaries in x, for $\beta x > 10$; this quantity is given by [12]

$$\left(\frac{dn_b}{dt}\right)_{3\rm b} = 0.90(3)^{9/2}\,\frac{n^3G^5m^5}{v_m^9}, \tag{6-37}$$

where n and v_m are the particle density and rms random velocity (three-dimensional), respectively, of the single stars. The numerical accuracy of the calculation leading to this expression is estimated as about six percent. Inserting numerical values we obtain

$$\left(\frac{dn_b}{dt}\right)_{3\rm b} = 1.91\times 10^{-13}\left(\frac{n}{10^4\ {\rm pc}^{-3}}\right)^3\left(\frac{m}{m_\odot}\right)^5\left(\frac{10\ {\rm km/s}}{v_m}\right)^9\ {\rm pc}^{-3}\,{\rm yr}^{-1}. \tag{6-38}$$

b. Tidal capture in two-body encounters

The heating effect in a self-gravitating system, as a result of tidal perturbations produced by a passing mass, has been discussed in §5.2c, where the impulsive approximation was used. The same effect occurs when two stars pass close to each other; the tidal perturbations produce accelerations of the stellar material relative to the stellar center. The internal energy gained by the stars is subtracted from the kinetic energy of their relative motion. While the internal energy ΔE_T gained by each star is a small fraction of the star's internal energy, it can be comparable with the kinetic energy of relative motion of the two stars at large separation. As before, we denote this initial kinetic energy by $m_r V^2/2$, where m_r is again the reduced mass for two passing stars. The two stars will capture each other in a bound orbit if

$$\Delta E_{T1} + \Delta E_{T2} \geqslant \tfrac{1}{2} m_r V^2, \tag{6-39}$$

where ΔE_{T1} and ΔE_{T2} are the internal energies gained by stars 1 and 2, respectively.

The elliptical orbit of the two stars immediately after capture will have an initial eccentricity only slightly less than unity. However, there will be many successive passages of the two stars at about the same periastron distance, and these will transfer further energy into the stars, which will presumably dissipate this small additional heating through a slight increase in radiant luminosity; the rate at which kinetic energy of internal motion gained in one periastron passage is dissipated into heat can influence the energy transfer in the next such passage, though it is too slow to affect the initial capture cross-section.

Ultimately the orbit should become nearly circular. Since the fraction of the initial angular momentum which is transferred to stellar rotation can scarcely exceed a few percent, because of the relatively small stellar radius of gyration, we assume that the orbital angular momentum remains constant. Hence in a circular orbit of radius R_1 and of relative velocity V_1, $R_1 V_1$ must equal the initial $R_p V_p$, where R_p is the initial periastron distance and V_p, the velocity at periastron. With V_p^2 obtained from the condition that $\frac{1}{2} m_r V_p^2 = Gm^2/R_p$ and V_1 from force balance in a circular orbit, one finds $R_1 = 2R_p$.

The value of ΔE_T and the resultant condition for tidal capture may be determined from the impulsive approximation. We use equation (5-37), which is based on the additional approximation that the relative motion of the two stars is rectilinear; we replace p and V by the periastron separation, r_p, and the relative velocity, V_p, at periastron. We reserve the symbol V to denote the relative velocity at infinity. While the impulsive approximation gives, in principle, an upper limit on the average energy transfer, the additional approximation of rectilinear motion introduces some error, though probably not a large one, into the final result. The quantities M_C and M_p in equation (5-37) are here replaced by m_1 and m_2. To simplify the analysis we assume that $m_1 = m_2 = m$, and that the two stars have identical radii, R_s.

If we combine equations (5-37) and (6-39), replacing m_r by $m/2$, we obtain as a condition for tidal capture,

$$r_p^4 < r_{pc}^4 \equiv \frac{32 r_m^2 G^2 m^2}{3V^2 V_p^2}. \tag{6-40}$$

We define r_{pc} as the critical value of the periastron distance, r_p, for tidal capture. The quantity r_m^2 is now the mean square radius for the material in the star. For a parabolic orbit $V_p^2 = 4mG/r_p$; if we introduce $v_{es}^2 = 2mG/R_s$, the escape velocity from the surface of one isolated star, equation (6-40) gives

$$\left(\frac{r_{pc}}{R_s}\right)^3 = \frac{4}{3}\left(\frac{r_m}{R_s}\right)^2 \left(\frac{v_{es}}{V}\right)^2. \tag{6-41}$$

For polytropes the ratio r_m^2/R_s^2, the mean square value of r/R_s, equals [13] 0.114 for $n = 3$ and 0.308 for $n = 1.5$. For stars with v_{es} equal to its solar value, critical values of (r_p/R_s) are given in Table 6.2, in the two columns headed "Impulsive." These are approximate upper limits, since any diminution in heating rate must be offset by a diminished r_{pc} to permit tidal capture.

As shown in chapter 5, the impulsive approximation is valid provided the frequency of the perturbation is not much less than the natural frequencies of the perturbed system. In the present case, the tidal perturbations excite

TABLE 6.2

Critical Periastron Distance for Tidal Capture

	Value of r_{pc}/R_s for			
V (km/s)	$n = 3$ Impulsive	$n = 1.5$ Impulsive	$n = 3$ Exact	$n = 1.5$ Exact
10	8.3	11.6	2.42	3.4
25	4.5	6.3	2.05	3.0
60	2.5	3.5	1.75	2.7

the non-radial modes of stellar oscillations. There are many of these, but for the modes with the fewest nodes, which are most effectively excited by the tidal perturbations, the angular frequency ω_m is usually some two to three times ω_s, defined as $(Gm/R_s^3)^{1/2}$. The angular frequency of the encounter, ω_p, is effectively V_p/r_p which equals $2\omega_s(R_s/r_p)^{3/2}$. Thus for $r_p/R_s = 2$, ω_p is about 0.3 times ω_m for the lowest modes, and the impulsive approximation may be a reasonable one. However, for r_p/R_s as great as 10, ω_p/ω_m is about 0.03, and the impulsive approximation is unreliable. For such slow changes, the shape of each star will slowly follow the changing form of the equipotential surfaces and net heating is markedly reduced.

To obtain more exact results, the excitation of the individual normal modes must be considered, and the heating summed over all modes. Analysis of this effect [14] has taken the parabolic motion of the two stars accurately into account, but in the tidal potential only the terms varying as $1/r^2$ and $1/r^3$ have been considered, an approximation valid for r_p/R_s larger than about 3. Detailed computations have been carried out for $n = 3$ [14,15] and for $n = 1.5$ [15], giving the results in Table 6.2, again for v_{es} equal to its solar value. As expected, the impulsive approximation is in rough agreement with the exact calculation for r_{pc}/R_s between 2 and 3, but the exact treatment gives a variation of r_{pc}/R_s about as $(v_{es}/V)^\mu$, where $\mu = 0.18$ for $n = 3$ and 0.12 for $n = 1.5$, instead of the more rapid variation given in equation (6-41), where $\mu = 2/3$.

The values of r_{pc}/R_s found in the more exact calculations are so small that the expansion used for the tidal potential is open to question. In particular, for $n = 3$ the stars are overlapping for $V = 60$ km/s and nearly so for $n = 1.5$. Since 90 percent of the mass for an $n = 3$ polytrope is within $R_s/2$ of the center, the overlap of the outer atmospheres may not seriously affect what happens to the stars as a whole.

If this difficulty is ignored, the exact results can be used to compute the rate of formation of close binaries. If we use the gravitational focussing equation (6-15)—with $m_1 + m_2 + m_3$ replaced by $m_1 + m_2 = 2m$ and $r_{\min}$ by r_{pc}—and average over a Maxwellian velocity distribution for V, then the rate of binary formation by tidal captures, which we denote by $(dn_b/dt)_{tc}$, is given by

$$\left(\frac{dn_b}{dt}\right)_{\rm tc} = \frac{1}{2} n^2 \langle \sigma V \rangle = 2\pi n^2 G m R_s \left\langle \frac{r_{pc}/R_s}{V} \right\rangle \qquad (6\text{-}42)$$

$$= 2\pi n^2 \gamma \frac{GmR_s}{v_m} (r_{pc}/R_s)_{10} \left(\frac{10^6 \text{ cm/sec}}{v_m} \right)^\mu,$$

where $(r_{pc}/R_s)_{10}$ denotes the value of this ratio for $V = 10$ km/s. The angle brackets denote an average over a Maxwellian distribution and the numerical constant γ, defined as $\langle (v_m/V)^{1+\mu} \rangle$, equals 1.01 and 1.00 for $n = 3$ ($\mu = 0.18$) and 1.5 ($\mu = 0.12$), respectively. As in equation (6-38), n is the number density of single stars, all assumed of identical mass m.

We now insert numerical values in equation (6-42). To determine $(r_{pc}/R_s)_{10}$ we take the values in Table 6.2 (computed for the solar value of v_{es}) and multiply by $(v_{es}/v_{es\odot})^\mu$, in accordance with the dependence of r_{pc}/R_s noted above for the exact calculation. Thus we obtain

$$\left(\frac{dn_b}{dt}\right)_{\rm tc} = 10^{-8}\, k \left(\frac{n}{10^4\ \mathrm{pc}^{-3}}\right)^2 \left(\frac{m}{m_\odot}\right)^{1+\mu/2} \left(\frac{R_s}{R_\odot}\right)^{1-\mu/2} \times \left(\frac{10\ \mathrm{km/s}}{v_m}\right)^{1+\mu} \mathrm{pc}^{-3}\ \mathrm{yr}^{-1}, \tag{6-43}$$

where $k = 1.52$ and 2.1 for the polytropes n equal to 3 and 1.5, respectively; v_m, here expressed in km/s, is again the rms velocity of the single stars, equal to $[\langle V^2 \rangle/2]^{1/2}$.

For degenerate stars and other compact objects colliding with normal stars of the same mass σ is reduced by $2^{-\mu/2}$ since energy is dissipated in only one star instead of two. Equation (6-43) remains valid if $n_d n_s$, the product of the two particle densities, replaces n^2, and k is increased by about $2^{1-\mu/2}$, becoming 2.9 and 3.9 for the polytropes n equal to 3 and 1.5, respectively. Values of k for stars of unequal masses have also been computed [15].

The binaries formed by tidal capture are generally very hard, with βx of order about 10^3. In contrast, binaries formed by three-body encounters are only marginally hard, with $\beta x \approx 10$. If the stars are non-degenerate and have solar values of m/R_s, a comparison of equations (6-38) and (6-43) shows that $(dn_b/dt)_{\rm tc}$ exceeds $(dn_b/dt)_{3b}$ unless n exceeds about $10^9\ \mathrm{pc}^{-3} \times (m_\odot/m)^3 \times [v_m/(10\ \mathrm{km/s})]^8$.

For the tidal captures included in these equations, r_p will be uniformly distributed from 0 to r_{pc}, since πp_1^2 is proportional to $r_{\min}$ in equation (6-15). For $r_p/R_s < 2$, two identical stars will experience a direct physical collision at least of their outer layers. The resultant processes are rather complex, with some of the heated gas escaping entirely. The stellar cores may preserve their integrity for a while, during a few orbits, and revolve about each other in a common envelope, finally coalescing into a single star of increased mass. Alternatively, as outer material is blown away by the collision and subsequent events, one or more compact objects may form. If tidal captures are important in the development of a cluster, as is indicated in chapter 7, such direct impacts must also be taken into account, since they comprise a large fraction of the encounters included in equation (6-42).

REFERENCES

1. D. C. Heggie, *M. N. Roy. Astron. Soc.*, **173**, 729, 1975.
2. P. Hut and J. N. Bahcall, *Ap. J.*, **268**, 319, 1983.
3. P. Hut, *Ap. J.*, **268**, 342, 1983.
4. P. Hut, *Ap. J. Suppl.*, **55**, 301, 1984.
5. P. Hut, *Ap. J.* (*Lett.*), **272**, L29, 1983.
6. J. G. Hills, *A. J.*, **80**, 809, 1975.
7. J. G. Hills and L. W. Fullerton, *A. J.*, **85**, 1281, 1980.
8. L. Spitzer and R. D. Mathieu, *Ap. J.*, **241**, 618, 1980.
9. R. S. Harrington, *Celest. Mech.*, **6**, 322, 1972.
10. S. Mikkola, *M. N. Roy. Astron. Soc.*, **207**, 115, 1984.
11. S. Mikkola, *M. N. Roy. Astron. Soc.*, **208**, 75, 1984.
12. J. Goodman and P. Hut, in preparation; quoted by P. Hut in *Dynamics of Star Clusters*, IAU Symp. No. 113, ed. J. Goodman and P. Hut (Reidel, Dordrecht), 1985, p. 231.
13. L. Motz, *Ap. J.*, **115**, 562, 1952.
14. W. H. Press and S. A. Teukolsky, *Ap. J.*, **213**, 183, 1977.
15. H. M. Lee and J. P. Ostriker, *Ap. J.*, **310**, 176, 1986.

7

Late Phases of Cluster Evolution

As the core of a cluster collapses, the rapid increase in central density accelerates the rate of binary formation and also the rate at which binaries harden. At the same time, the heavier stars become more and more concentrated within the core, while the core mass steadily decreases. In addition, the granularity of the gravitational potential becomes progressively more and more important, with the gravitational force of the nearest neighbor becoming comparable with the force produced by the inner stars; the relative effect of close encounters steadily increases and the Fokker-Planck approach becomes less accurate. If stars of normal radii are present, close tidal interactions and direct collisions also occur in the collapsing core. Evidently the physical processes occurring in this evolutionary phase show marked differences from those discussed in chapters 2 through 5 above.

This final chapter discusses some of these processes and analyzes their effect on cluster evolution. Since binary stars provide a major source of energy, and are relatively well understood, the first section analyzes the rate at which this power is delivered in a cluster and where the energy goes. In this discussion the inner regions of the cluster are represented by an isothermal sphere. Most model clusters show nearly isothermal conditions within the inner half of the cluster's mass; i.e., for $r < r_h$. Even during gravothermal collapse the structure is close to isothermal for $r/r_c \leqslant 10$, as pointed out in §3.3b. Section 7.2 treats the termination of the collapse phase as a result of the energy given up by one or more hard binaries. In §7.3 subsequent reexpansion of the cluster's inner regions is discussed. Since there may be many complications involving collisions with normal stars, and since we do not know the abundances of such stars relative to the compact objects—white dwarfs, neutron stars and black holes—this final section and indeed the entire post-collapse phase of the cluster are both somewhat uncertain.

7.1 BINARIES AS AN ENERGY SOURCE IN CLUSTERS

To help understand the effects associated with binary stars we give here a few basic results which follow from applying the previous chapters to the inner isothermal regions in a cluster. First we consider the rate of binary formation per unit relaxation time. This quantity is of importance since it is the relaxation time that determines the evolution rate. As shown in chapter 4, the core collapses in a time less than about some $20t_{rh}$, with shorter times for systems with a distribution of masses; in the core at a time t the time remaining until collapse is several hundred times t_{rc}, the relaxation time at the center, evaluated at the time t—see equation (4-17). Next the rate of energy liberated by binaries, per unit relaxation time and per unit volume, is compared with the local density of kinetic energy. The distribution of this liberated energy over different regions of the cluster is subsequently analyzed.

We consider first the rate at which binaries are formed by tidal capture. We denote here by N_b the total number of such binaries in the cluster; to compute dN_b/dt, we multiply equation (6-42) by $4\pi r^2\, dr$ and integrate over the cluster. From the tabulated properties [1] of an isothermal sphere we find that the integral of $n^2\, dV$ over the system equals $1.07 n_c N_c$; the core region, whose radius r_c we here set equal to 3κ, contributes about half the total integral. Similarly we find that N_c, the number of single stars within the radius r_c, equals $0.517 \times 4\pi n_c r_c^3/3$. If the stars which capture each other are assumed to be polytropes of index 1.5, an appropriate assumption for the main-sequence stars, we find

$$\frac{dN_b}{dt} = 7.2\pi n_c N_c G m R_s \frac{1}{v_m}\left(\frac{10^6 \text{ cm/s}}{v_m}\right)^{0.12}, \tag{7-1}$$

where n_c is the value of n at $r = 0$ and we have inserted the value of $(r_{pc}/R_s)_{10}$ from Table 6.2 (p. 144), ignoring the correction factor $(v_{es}/v_{es\odot})^{0.12}$; we omit the subscript c from $v_{mc} = v_m(0)$. As before we let n_c as well as N_c refer to single stars, and assume that these quantities much exceed the corresponding values for binaries.

If now we multiply equation (7-1) by t_{rc}, the relaxation time at the center, using equations (2-61) and (2-13), equation (7-1) yields

$$t_{rc}\frac{dN_b}{dt} = \frac{2.9 N_c}{\ln \Lambda}\left(\frac{v_m}{v_{es}}\right)^2 \left(\frac{10 \text{ km/s}}{v_m}\right)^{0.12}. \tag{7-2}$$

In equation (2-14) $\ln \Lambda$ was evaluated for conditions inside the radius r_h. In the present case, where we are considering the relaxation time at the cluster center, the appropriate value of $p_{\max}$ is r_c; at this distance from the center n/n_c has fallen to 0.35. If we express p_0 with use of equation (2-5), replacing V^2 by $2v_m^2$, and eliminate v_m^2 with use of equation (1-24), we obtain in place of equation (2-14)

$$\ln \Lambda = \ln\left(\frac{4\pi n_c r_c^3}{3}\right) = \ln(1.9 N_c). \tag{7-3}$$

If we set v_m equal to 20 km/s, take the solar value of 617 km/s for v_{es}, and let $\ln \Lambda = 8$, its value for $N_c \approx 10^3$, equation (7-2) gives $3.5 \times 10^{-4}\, N_c$ for

$t_{rc}\, dN_b/dt$. During a time interval of $100t_{rc}$, about the time interval required for appreciable evolution, N_b will increase by about 0.04 N_c.

If we assume that core contraction occurs in accordance with the gravothermal instability, then t_{rc} and N_c will vary as $t_{\text{coll}} - t$ and $(t_{\text{coll}} - t)^{0.42}$, respectively, in accordance with equations (3-9) and (3-6). Equation (7-2) can be integrated directly; with $t_{\text{coll}} = 190t_{rc}(0)$, obtained from equation (4-17) for the realistic anisotropic velocity distribution, and with the other numerical values used above we obtain

$$N_b(t) = 0.16(N_c(0) - N_c(t)). \tag{7-4}$$

In obtaining this approximate result we have assumed that $N_b(0)$ vanishes and we have ignored the variations of v_m and $\ln \Lambda$ with time. Thus as the core shrinks, N_b increases, and equals $N_c(t)/2$, corresponding to a mass in binaries equal to that in the core, when N_c has fallen to about a fourth of its initial value, corresponding to an increase of ρ_c by about a factor 50. The assumption that n_c is unchanged by the formation of binaries breaks down before this point is reached, and tidal capture presumably occurs before the gravothermal instability, with its accompanying functional forms for t_{rc} and N_c, is well developed. Hence equation (7-4) is illustrative only, and shows that the tidal capture process will convert a large fraction of core stars into binary systems relatively early in the collapse phase.

Next we consider the corresponding rate at which hard binaries form by three-body encounters. We now use N_b to denote the total number of binaries formed by three-body processes. Again we integrate dn_b/dt, this time obtained from equation (6-37), over the volume of an isothermal sphere to determine dN_b/dt. In this case about 80 percent of the binary formation occurs within the core radius; the integral of $n^3\, dV$ over the system equals $0.42n_c^2 N_c$. If we multiply dN_b/dt by t_{rc}, and introduce again the quantity N_c, the total number of stars within the radius r_c, we obtain after some algebra

$$t_{rc}\frac{dN_b}{dt} = \frac{0.22}{N_c \ln \Lambda}. \tag{7-5}$$

This equation can be integrated over dt, with the variations of $t_{rc}(t)$ and $N_c(t)$ again taken from equations (3-9) and (3-6). If N_b is assumed to vanish initially and the change of $\ln \Lambda$ with time is again ignored, we obtain

$$N_b(t) = \frac{100}{\ln \Lambda}\left(\frac{1}{N_c(t)} - \frac{1}{N_c(0)}\right). \tag{7-6}$$

As before, we have let $t_{\text{coll}} = 190t_{rc}(0)$ from equation (4-17), corresponding to $f = f(E,J)$.

If we use equation (7-6) somewhat arbitrarily for N_b as small as unity but with $N_c \ll N_c(0)$, and take $\ln \Lambda$ from equation (7-3), we find that the first binary appears when N_c is 25. This result is clearly very approximate. Quite apart from the use of equation (7-6) for the first binary to be formed, the assumptions made in deriving dN_b/dt are inaccurate when N_c is so small. For example, a soft binary of semi-major axis a in a core of uniform particle density n_c will be subject to a tidal force exceeding Gm^2/a^2 if a exceeds $(4\pi n_c/3)^{-1/3}$, about the mean distance from a star to its nearest neighbor; in an isothermal core the corresponding $\beta x \approx N_c^{-2/3}$, or about 0.1 when $N_c = 30$. This tidal force is entirely compressive rather than disruptive but will strongly affect the dynamical processes involving soft binaries with βx less than this value. According to the analysis cited in §6.3a, about 40 percent of the hard binaries formed in three-body encounters originate as soft binaries with $\beta x < 0.1$. Hence environmental perturbations of soft binaries will presumably change the details of binary formation, though it seems likely that equation (7-6) gives at least the correct order of magnitude.

We consider next the liberation of energy by hard binaries, and express the rate for this process in an intuitively simple form. The average rate dx/dt at which a binary gives up energy and hardens correspondingly is given by equation (6-30). If we multiply this by $n_b t_r$ we obtain ΔU_b, the mean energy released per unit volume per local relaxation time. Dividing by U_s, the local density of kinetic energy of single stars, equal to $n_s m v_{m}^2/2$, yields, with use of equations (2-61) and (2-13),

$$\frac{\Delta U_b}{U_s} = \frac{0.92}{\ln \Lambda} \times \frac{n_b}{n_s}. \qquad (7\text{-}7)$$

In obtaining this result we have set $A = 21$ in equation (6-30) and also $\beta_r = \beta_s$. Evidently if n_b/n_s is a few percent, the energy released within a hundred relaxation times will about equal the kinetic energy. If binary-binary reactions are taken into account the energy liberated will be several times greater. Equation (7-7) is valid for binaries of any hardness, but as we shall see below the energy released by the very hard tidal-capture binaries goes mainly into kinetic energy of escaping reaction products rather than into the cluster.

Equation (7-7) is somewhat incomplete in that it neglects heating energy associated with binary formation. For the formation of binaries by three-body encounters this heating effect is appreciable, and can be taken into account approximately by including in n_b all binaries with $\beta x \geqslant 3$; for this extended range in βx the survival probability exceeds 0.5. In the tidal capture process translational kinetic energy is lost rather than gained. On the average half of this kinetic energy is associated with the relative motion of the two stars, and equals $\frac{1}{2}m_r V^2$. When these stars become gravitationally bound, this energy is all converted by tidal interactions into internal stellar energy; the cluster energy, computed with each single star and each binary replaced by a mass point, decreases.

The evolutionary effects of the binary energy released in accordance with equation (7-7) naturally depend on where this energy is deposited. If the change $y \equiv \Delta x$ in the binary binding energy x is not much greater than $\frac{1}{2}mv_m^2$, the mean kinetic energy of the single stars, the orbits of the stars which gain this energy will not be grossly affected, and the energy will be locally deposited. In this case, the binary produces "kinetic heating" of the inner cluster. If y appreciably exceeds $\frac{1}{2}mv_m^2$, a star which gains such energy will move far out in the cluster; since energy will be released from binaries mostly in the central regions, such energetic stars will be moving in nearly radial orbits, and as we shall see below much of this energy will be reabsorbed elsewhere in the cluster rather than in the core.

Finally, if y exceeds 1.5 $m\phi(0)$, the star will on the average receive two-thirds of this energy and will escape from the cluster; if y exceeds $3m\phi(0)$ the accelerated binary is also likely to escape. As pointed out in §4.2c, the dimensionless potential X_0 increases from 9 up to 13 as ρ_c increases by a factor of 10^6 during gravothermal collapse; hence if we use equations (6-8) and (4-15) for β, we see that all the reaction products are likely to escape if $\beta\,\Delta x$ exceeds 40. If we set $\langle \Delta x \rangle / x \approx 0.4$ from equation (6-25), we find that escape of most of the reaction products is likely if $\beta x > 100$. Tidal-capture binaries tend to be harder than this, and much of the energy liberated will be carried away from the cluster by energetic reaction products. This conclusion is strengthened by the greater energy released in binary-binary encounters. When the reaction products escape, the only energy gained by the cluster is the decrease in gravitational energy resulting from the decrease of mass in the bound system. This process we call "gravitational heating" of the cluster.

We consider here first the effective distribution of kinetic energy which is imparted to energetic but bound reaction products and, second, the distribution of gravitational energy which the cluster gains from escapers. As above, the discussion is restricted to the inner isothermal region of the cluster.

An energetic reaction product, which for simplicity we regard as a single star, is assumed to move in an initially radial orbit. Encounters with other stars along its orbit will result in an energy loss, produced mostly by dynamical friction. If the orbit remained fixed, most of the initial kinetic energy, which we denote by E_0 per unit mass, would be absorbed in the high-density core during successive orbits. However, the velocity perturbations at large r, near the apocenter of the orbit, will increase J, the angular momentum per unit mass. Since $J = r_p v_p$, where v_p is the velocity at pericenter, at which $r = r_p$, an increase in J increases r_p; v_p is not much affected, and may be set equal to its initial value $(2E_0)^{1/2}$. We shall show that when E has been decreased well below E_0, r_p has increased to a value substantially exceeding the core radius, r_c. Hence the initial kinetic energy of an energetic reaction product tends to be imparted to cluster regions well outside the core.

To demonstrate this result, we use the orbital-averaged diffusion coefficients introduced in §2.2b. The quantity $\langle \Delta E \rangle_{\text{orb}}$ denotes the mean change of E per unit time, averaged over the orbit as in equation (2-91). Similarly, the mean square change of J per unit time, equal to v_p^2 times the mean square change of r_p, equals $\langle r^2 (\Delta v_t)^2 \rangle_{\text{orb}}$, where Δv_t is the increment of transverse velocity, perpendicular to $\mathbf{r}$; for the nearly radial orbits we consider, $\mathbf{v}_t$ is perpendicular to the stellar velocity, and $\Delta v_t = \Delta v_\perp$. We define $-Y$ as the ratio of this second orbital-average diffusion coefficient to the first, both evaluated for the initial orbit of the energetic reaction product; evidently Y is the initial mean square change of J per unit mean decrease of E.

In evaluating the orbit-average diffusion coefficients we may approximate their dependence on velocity. The mean energy loss is produced primarily at small r, where the velocity of the energetic reaction product substantially exceeds v_m. Hence in equation (2-55) we take the asymptotic form for large x. Because of the factor r^2, the mean square change in J is produced mainly near apocenter, where $v \ll v_m$. As a result, in equation (2-54) we take the form for small x, using equations (2-58). After some algebra we obtain for the ratio Y,

$$Y \equiv \frac{\langle (\Delta J)^2 \rangle_{\text{orb}}}{-\langle \Delta E \rangle_{\text{orb}}} = -v_p^2 \frac{d\langle r_p^2 \rangle}{d\langle E \rangle}$$
$$= 4\left(\frac{2}{3\pi}\right)^{1/2} \frac{1}{v_m} \int_0^{r_a} \frac{\rho r^2\, dr}{v_r} \bigg/ \int_0^{r_a} \frac{\rho\, dr}{v_r^2}. \tag{7-8}$$

To evaluate the integral in the denominator we set v_r^2 equal to v_p^2, its value at $r = 0$, which in turn is about equal initially to $2E_0$, where E_0 is the initial kinetic energy of the reaction product. Thus we neglect the change of v_r^2 as the star moves through the central region. The integral may then be evaluated numerically for an isothermal sphere [1], yielding

$$\int_0^\infty \rho(r)\, dr = 3.0 \kappa \rho(0). \tag{7-9}$$

In the other integral we assume that r/κ is so large that we may use for ρ and for the dimensionless potential θ the asymptotic expansions in equations (1-29) and (1-28). The product ρr^2 is then constant, and with $v = v_r$ we find from equations (1-6), (1-22) and (1-25)

$$v_r = (\tfrac{2}{3})^{1/2} v_m (\theta_a - \theta)^{1/2}, \tag{7-10}$$

where θ_a denotes the value of $\theta(\xi)$ at apocenter, $\xi = r_a/\kappa$. With dr replaced by $(dr/d\theta)\,d\theta = 2^{-1/2}\kappa e^{\theta/2}\,d\theta$, the integral can then be evaluated exactly. Equation (7-8) becomes

$$\frac{d\langle r_p^2 \rangle}{-d\langle E \rangle} = \frac{4(2)^{1/2}}{3v_m^2}\,\kappa r_a. \tag{7-11}$$

To compute $\langle r_p^2 \rangle$ when the initial kinetic energy has all been lost, we assume that the right-hand side of equation (7-11) is constant as r_a and v_p^2 change in successive orbits. This should give an approximate value for $\langle \Delta r_p^2 \rangle$ when $-\Delta E$ is about equal to E_0. With $r_p = 0$ initially we obtain

$$\frac{\langle r_p^2 \rangle}{r_c^2} = 0.31\,\frac{v_p^2}{v_m^2}\,\frac{r_a}{r_c}. \tag{7-12}$$

Both v_p^2/v_m^2 and r_a/r_c in this equation substantially exceed unity. If we let v_{ce} equal the escape velocity from the center, the ratio $v_{ce}^2/v_m^2 = 2X_0/3$ varies—see §4.2c—between 6 and 9 during gravothermal collapse. For energetic bound particles this ratio is smaller but still appreciable. The corresponding value of r_a/r_c can be large; for example, r_h/r_c at the start of gravothermal collapse is at least 10 and increases by several orders of magnitude as the collapse proceeds. Thus $\langle r_p^2 \rangle$ substantially exceeds r_c^2 by the time that the energetic particle has been slowed down.

This discussion shows that only a fraction of the energy carried off by the energetic reaction products is transmitted by dynamical friction to stars within the core. In addition, at late stages of the collapse, when the increased central density gives a time τ to collapse—in equation (4-17)—appreciably less than t_{dh}—see equation (4-1)—most of the reaction products will not return to pericenter in time to affect the collapse. We conclude that direct kinetic heating of the core by such energetic stars and binaries, with $r_a \gg r_c$, has only a slight effect on core collapse.

The other mechanism for transferring energy from binaries to the cluster is through a change in W, the gravitational self-energy of the cluster. When a star or binary of mass m at an initial radius r_i achieves an energy sufficient to escape from a cluster, the work which it does against the cluster's gravitational field is $m(\phi(\infty) - \phi(r_i))$; for simplicity we assume that the cluster is isolated and that $\phi(\infty)$, the gravitational potential at infinity, is zero. Since the escaping particle has zero gravitational energy at infinity, this work done must increase W by an amount $-m\phi(r_i)$.

It is possible to express this change in energy in terms of $\delta\phi(r)$, the change of potential in the cluster at each r resulting from ejection of a mass δm_j

initially at radius r_j. The gravitational energy W of the cluster is given by

$$W = \tfrac{1}{2}\sum m_j \phi(r_j), \tag{7-13}$$

summed over all masses m_j in the cluster. The factor 1/2 is present because ϕ varies as the total mass M if the cluster is gradually disassembled, without change in the smoothed $\rho(r)/\rho(0)$. From equation (7-13) it follows that if the mass of one spherical shell, at radius r_i, is altered by δm_i (also spherical), we obtain to first order

$$\delta W = \frac{1}{2}\,\phi(r_i)\,\delta m_i + \sum_j \frac{1}{2}\,m_j\,\delta\phi(r_j), \tag{7-14}$$

where

$$\delta\phi(r_j) = G\delta m_i \times \begin{cases} 1/r_i & \text{if } r_i > r_j, \\ 1/r_j & \text{if } r_i < r_j. \end{cases} \tag{7-15}$$

If the sum in equation (7-14) is converted to an integral, an integration by parts indicate that this second term equals the first term, giving $\delta W = \delta m_i \phi(r_i)$, as required.

This discussion suggests that the gravitational energy per unit mass at r_j increases by $\delta\phi(r_j)/2$. Alternatively, if attention is focussed on the work required to lift a particular spherical mass element out of the cluster, without changing other masses, this energy per unit mass clearly changes by twice as much. This uncertainty in assignment of gravitational energy to specific masses results from the fact that gravitational energy, unlike kinetic energy, is not a localized phenomenon, except in the special case of a fixed gravitational field.

When a star or binary is ejected, the kinetic energy that it had before it was accelerated is lost to the cluster, decreasing the total kinetic energy T and somewhat offsetting the increase in W. Unlike the heating produced by ΔW, the cooling associated with ΔT is localized.

To indicate the net effects expected from gravitational heating of the cluster, we examine this process more carefully. To obtain results that are both precise and simple, we compute the initial acceleration resulting when mass is abruptly removed from the central region of a cluster. We assume that initially the density distribution in this region is that of an isothermal sphere, and that at the time $t = 0$ the density ρ is reduced by an amount $\delta\rho$, which at each radius we assume is proportional to $\lambda\rho^3$, where λ is a small constant. This mass-loss rate is appropriate for the interaction of binaries with single stars, if the ratio of binaries to single stars is small even in the core and if equipartition of translational kinetic energy exists between these two components; under these conditions ρ_b varies as $\rho_s^2 \approx \rho^2$. The mass loss is assumed to have no direct effect on v_{ms}, the mean square stellar velocity.

We idealize the problem by assuming a perfect gas, initially in hydrostatic equilibrium; then for the acceleration in the r direction, we have

$$\frac{du_r}{dt} = -\frac{v_m^2}{3}\,\delta\left(\frac{1}{\rho}\frac{d\rho}{dr}\right) - \frac{4\pi G}{r^2}\int_0^r r_1^2\,\delta\rho\,dr_1, \tag{7-16}$$

where a subscript 1 is used to denote the variable of integration. We assume

$$\delta\left(\frac{\rho}{\rho_c}\right) = -\lambda\left(\frac{\rho}{\rho_c}\right)^3, \tag{7-17}$$

where ρ_c, the initial central density, is a constant and ρ on the right-hand side refers to values in the zero-order equilibrium; we consider only terms to first order in λ, with $u_r = \lambda\psi v_m$. Introducing as variables the dimensionless zero-order density $w \equiv \rho/\rho_c$ and the dimensionless radius $\xi \equiv r/\kappa$—see equation (1-24)—we find

$$\frac{\partial\psi}{\partial\tau} = 2w\frac{dw}{d\xi} + \frac{1}{\xi^2}\int_0^\xi w^3\xi_1^2\,d\xi_1, \tag{7-18}$$

where

$$\frac{d\tau}{dt} \equiv \left(\frac{4\pi G\rho_c}{3}\right)^{1/2}. \tag{7-19}$$

For small ξ, equation (7-18) may be evaluated from the series solution in equation (1-27). Computation of $\partial\psi/\partial\tau$ from the known values [1] of w gives the results in Fig. 7.1, plotted against ξ. The core radius is at $\xi = 3$. If λ is positive, corresponding to loss of mass, the inner part of the core contracts, with $\partial\psi/\partial\tau$ negative, despite the fact that $\delta\phi$ resulting from mass loss has its maximum value there. Expansion occurs at greater radius, at ξ exceeding about 2.4 (i.e., $r > 0.8r_c$), outside the region where most of the mass loss occurs. It is not surprising that mass loss from the core tends to produce an inward velocity to equalize the pressure. A theoretical consequence of Fig. 7.1 is that gravitational heating of the core does not seem to inhibit core collapse directly. In these idealized models it is the expansion of the regions just outside the core which indirectly can lead, in a higher approximation, to termination of the collapse.

This analysis and the resultant Fig. 7.1 remain valid if the mean free path much exceeds r_c, provided that the inner cluster regions are initially in isothermal equilibrium. In this situation, mass loss will decrease the inward acceleration of all stars, but for $\xi < 2.4$ the relative lack of stars will lead in a short time to a net inward stellar flux across each spherical shell; the corresponding average u_r for the stars in a unit volume will be inward. The subsequent evolution of the core and the cluster as a whole may well depend

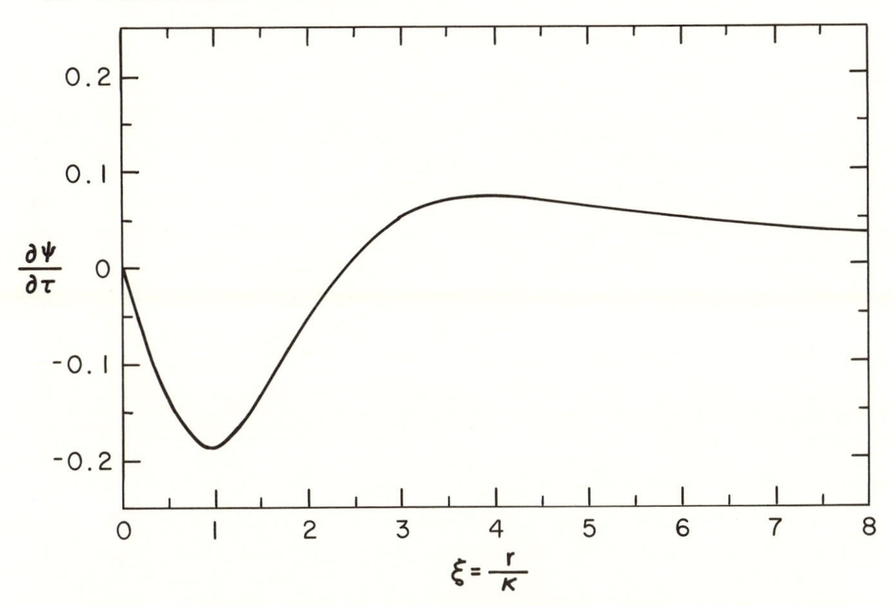

Fig. 7.1. Initial Acceleration Produced by Gravitational Heating. The curve shows the initial acceleration, in dimensionless units, when mass is abruptly removed from the central region, the loss rate varying as $[\rho(r)]^3$. The horizontal scale is the dimensionless radius $\xi \equiv r/\kappa$.

on the magnitude of the mean free path. In any case mass loss certainly decreases the binding energy of the cluster, but the detailed effects on different regions of the cluster, especially on the core itself, are not necessarily simple.

7.2 TERMINATION OF COLLAPSE

Detailed model calculations have shown that binary stars, formed in the central regions of a cluster with a collapsing core, can in fact terminate core collapse. This conclusion has often been assumed, ever since theory has revealed collapsing cores in evolving globular clusters. To establish this result firmly is of substantial importance. Here we summarize the relevant calculations, first for the termination of collapse by newly formed three-body binaries and then, by newly formed tidal-capture binaries. The presence of primordial binaries could presumably also terminate the collapse at some stage, but the calculations so far carried out [2] do not show this effect, perhaps because these models did not extend so very far into the collapse phase.

a. Binaries formed in three-body encounters

We have seen above that to form even one binary in a collapsing core requires a core containing some 25 stars if the equations derived for large N are extrapolated to such low N_c values. To analyze the evolution of such cores, the Fokker-Planck equation is unsuitable. The one method that carries assurance in this problem is the use of detailed dynamical integrations, following the orbit of each star numerically. Practical limits on computing prevent such calculations for all the stars in a globular cluster.

To analyze the evolution of such a cluster in the core-collapse phase, a hybrid program has been devised [3], which combines dynamical N-body calculations for stars in the central region of the cluster, at radii less than some particular value r_N, with a Fokker-Planck approach for stars at larger r. The primary change at the transition radius r_N is in the method used for computing the potential. For $r < r_N$ the potential gradient is computed exactly, considering the contribution from all stars within r_N, but ignoring those with $r > r_N$. This inner zone where N-body integration techniques are used is called the NB region. For $r > r_N$ the potential is assumed to be spherical, with a gradient equal to $GmN(r)/r^2$, where $N(r)$ is the number of stars at distances less than r from the cluster center. The granularity of the gravitational field is taken into account in this outer region by a Fokker-Planck technique; this outer zone is called the FP region.

When a star is outside the radius r_N, (but within a larger radius, discussed below), the integration method used is that of the dynamical Monte Carlo computations (see §4.1a), with r obtained by integrating equation (4-3). As in the Princeton Monte Carlo models, the perturbations produced by random stellar encounters are calculated separately, and applied at intervals to the stars in this region, with velocities adjusted to ensure that energy is conserved in this process. Among differences in detail from the Princeton computations are: (1) the perturbations are computed in E and J rather than in v_r and v_t; (2) instead of applying these perturbations to each star at short time intervals, diffusion coefficients are integrated along the unperturbed orbit and applied to each star at substantial intervals—when a star crosses r_N or passes through apocenter or when the cumulative perturbations become appreciable.

Another feature in the FP region is that when a star moves outward across the radius r_N, its angular position is retained and updated, so that when it recrosses r_N it will reenter the NB region in the correct position. To this end, the equation for $d\theta/dt$ in the orbital plane is integrated. The orientation of this plane is also conserved during the integration, and modified by perturbations which give velocities perpendicular to the initial orbital plane.

While dynamical integrations with a spherically symmetrical potential are much less demanding of computer time than exact N-body computations, they are more demanding than the Fokker-Planck (FP) models which use orbit-averaged perturbations of E and J. As indicated in §§4.1b and c, such FP models use a time step which is small compared to the relevant relaxation time but large compared to the dynamical crossing time. To take advantage of this feature in the hybrid code developed for clusters, the FP region is subdivided, with dynamical integrations used in the intermediate FP-1 region, while stars whose orbits lie entirely in the outer FP-2 zone are treated entirely in terms of $f(E,J)$. This outer region is defined by $r > Kr_N$; in the computation reported here, K was set equal to 3.

For a star which crosses Kr_N as it moves outward, information on position and orientation of the orbital plane is discarded, with the number of FP-2 stars in the appropriate E,J bin increased by 1. However, for each such star the time spent in this outer region is computed, so that the star can be reinserted in the intermediate FP-1 region at the proper time, with random choice of angular position and orbit orientation. Changes in $f(E,J)$ resulting from two-body encounters in the FP-2 region are taken into account. However in the computations leading to the results reported here, the spherical potential was taken to be independent of time in FP-2, an approximation consistent with the small changes found for $f(E,J)$ within this region.

This hybrid code has been applied [3] to a cluster in the late collapse stage, with all stars of identical mass. In the initial state, $\rho(r)$ and $v_m(r)$ were taken from the self-similar solution described in §3.3b and confirmed for large N by detailed models—see §4.2c. The initial number of stars was taken to be 220 in FP-1 and about 100 in the inner NB region, with N_c, the number of stars in the core, equal to 55 initially; the corresponding core radius r_c equalled $0.6r_N$. The evolution was followed during an interval equal to about 270 times the initial value of the central relaxation time t_{rc}, computed as usual from equations (2-61) and (2-13), and modified here to take equation (7-3) into account; according to this equation, for N_c equal to its mean value of 40 during the collapse computations, $\ln \Lambda = 4.3$ instead of the value 2.8 obtained from equation (2-14).

A principal result is that the homologous collapse assumed prior to the start of the integration continues with no apparent change as long as no hard binaries are formed. The lower plot b in Fig. 7.2 shows the values of t_{rc} obtained from the calculations. Both t_{rc} and t are given in units of an initial dynamical time t_N. The straight line provides a good fit with the data for $t < 500t_N$. This observed linear decrease of t_{rc} with the elapsed time t is in close agreement with equation (3-9). During this same interval, for t between 0 and $500t_N$, the changes of r_c, v_{mc} and n_c are also consistent with equations (3-5) through (3-8), if ζ is again set equal to 0.737. It is somewhat remarkable that the same self-similar solution is valid when N_c is so small, decreasing from 55 initially to about 30 when the collapse stops.

A second important result is that the collapse stops as soon as a hard binary appears. The upper plot *a* in Fig. 7.2 shows the binding energy of the one hard binary present during this period. While marginally hard binaries ($\beta x = 1.5x/\frac{1}{2}mv_m^2 \approx 1$) appear and disappear for t/t_N between 450 and 500, a binary with $\beta x \approx 4$ appears at $t/t_N \approx 500$, and one time unit later increases in hardness to $\beta x \approx 14$, with more gradual increases subsequently. The lower plot shows that t_{rc} stops its decrease at about this time, when r_c turns around and starts to increase, with n_c decreasing. The total increase of binding energy for this one binary, as t/t_N increases 510 to 610, is 40 times the initial value of $\frac{1}{2}mv_n^2$, or about 30 times the value of this quantity when the binary is formed. Since there are about 30 stars within the core, the

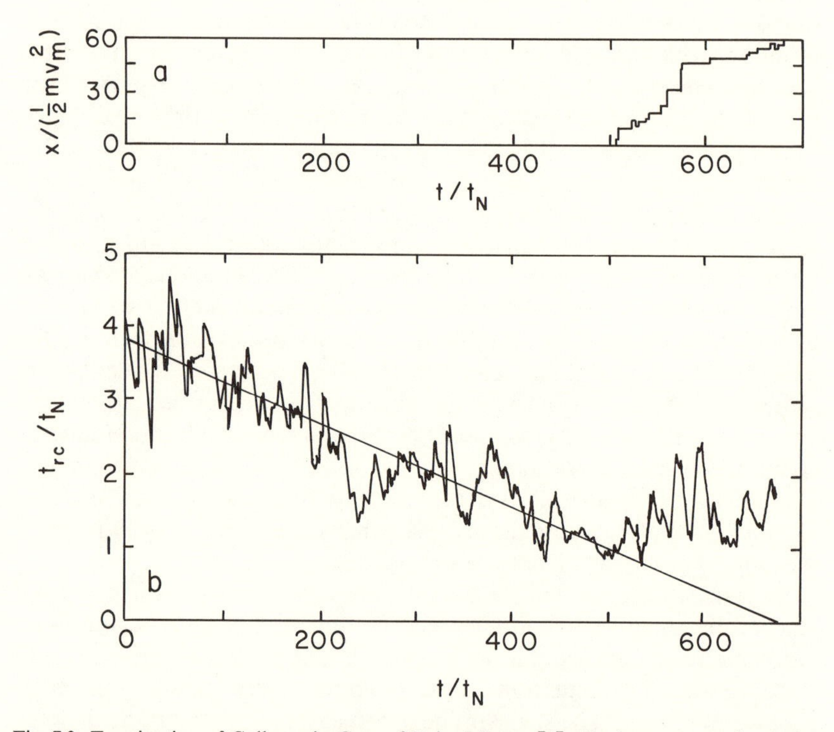

Fig. 7.2. Termination of Collapse in Core of Point Masses [3]. The lower curve shows the central relaxation time t_{rc} in units of the dynamical time t_N. The upper curve shows the binding energy x of the hard binary which appears at about $500t_N$; x is divided by the mean kinetic energy at $t = 0$.

energy given up by the binary during $100t_N$ is just enough to double the kinetic energy of each star in the core. Since the energy liberated is mostly locally deposited, the termination of core collapse is to be expected.

This heating rate may be compared with that obtained from equation (7-7). With $\ln \Lambda$ set equal to 4.3 and $n_b/n_s = 1/30$, this equation gives $\Delta U_b/U_s = 7.1 \times 10^{-3}$ in one relaxation time. Since $t_{rc} = t_N$ when the binary is formed, the predicted value of $\Delta U_b/U_s$ within a time interval of $100t_N$ is 0.7, if we neglect the small changes of n and v_m^2 during time interval. This result may be compared with the value 1.0 found above from Fig. 7.2a. This comparison is independent of the value of $\ln \Lambda$ used; since $\Delta U_b = t_{rc}\, dU_b/dt$ and t_{rc} varies inversely with $\ln \Lambda$, the $\ln \Lambda$ factor on the left-hand side of equation (7-7) cancels that on the right-hand side. Considering the statistical fluctuations which occur in binary hardening, this agreement is about as close as might be expected.

The fact that the gravothermal collapse continues even for very small N_c encourages one to compare the observed collapse rate with that obtained for the large-N models. Fig. 7.2 shows that at any moment the extrapolated time to collapse is equal to 180 times the value of t_{rc} at that moment. This result and the corresponding value of 6×10^{-3} for ξ_c are in agreement with the results given in equations (4-17) and (4-16) for collapsing models with velocity diffusion in E,J space. So close an agreement with the anisotropic velocity diffusion models may be a chance coincidence in view of possible errors in these various models.

In any case, close agreement between the Fokker-Planck models and precise calculations for small N is scarcely to be expected. When N is small, the assumption of many small velocity changes, which is basic to the Fokker-Planck equation, as we have seen in §2.1a, becomes unrealistic. In discussing any process dependent on relaxation, an effective value of $\ln \Lambda$ can be introduced to give the correct results, but the theoretical prediction of such an effective value is not yet possible. A detailed calculation [4] of the diffusion coefficient $\langle(\Delta v)^2\rangle$ shows that the variation of the logarithmic term in the integration over V, the relative velocity, reduces the argument of $\ln \Lambda$ in this particular case by an additional factor 0.4. However, higher-order terms in the Fokker-Planck equation presumably introduce comparable corrections. Thus the constant factor in Λ is uncertain. For evaluation of t_{rc}, equation (7-3) has some advantage in that it gives the agreement of collapse rates noted above between the hybrid model and those of §4.2.

The rate of binary formation is consistent with the theory leading to equation (7-6). In the discussion of that equation, appearance of the first hard binary was predicted when N_c fell to 25; in the hybrid model such a binary appeared at $N_c = 30$. As pointed out in §7.1, equation (7-6) is approximate, but at least it is unaffected by uncertainties associated with the theoretical value of ln Λ and hence of t_{rc}.

An apparent discrepancy between theory and the hybrid model is that the first hard binary increases its hardness so abruptly; βx increases from 4 to 14 almost immediately after the binary appears. Theory would predict a slower growth, with a number of encounters normally required to increase βx to a value where dissociation is improbable. However, in several other runs the first hard binary to appear shows a more gradual increase in x, suggesting that the abrupt early increase in βx shown in Fig. 7.2a is an unusual statistical fluctuation.

At $t \approx 630 t_N$ the binary responsible for terminating core collapse in this particular run captures a star and becomes a long-lived triple system; at later times this disrupts, with both reaction products ejected from the core, and is effectively lost. The core collapse resumes, but is soon stopped as other hard binaries are formed.

This model indicates that three-body encounters can terminate core collapse. Apart from the assumption of equal masses for all stars, and their treatment as mass points, the model seems free from major approximations or simplifications and the results appear reliable. In an actual cluster the presence of stars with different masses may modify the details of this process. In particular, if a core of some 40 compact objects contains a few relatively heavy ones, these more massive objects will almost certainly exchange with the lighter ones in any binary formed. A binary composed of two massive compact objects will give up energy to lighter objects, with a reduced probability of itself being ejected. In contrast, the presence of two such heavy binaries would be transitory. In an encounter between them one of the two binaries would likely be disrupted, with its components ejected. If there are any black holes present in the cluster, one massive binary formed of the two heaviest such objects could be a relatively permanent feature of the cluster core, and would provide an energy source for continuing expansion of the system.

b. Binaries formed by tidal capture

The binaries formed by mutual tidal capture of normal stars differ in two major respects from those formed by three-body encounters. First, they are formed much more rapidly for typical cluster conditions, as is evident from a comparison of equations (6-38) and (6-43). Second, their heating efficiency is less, since reaction products produced by very hard binaries have kinetic energies much greater than the escape energy. We consider the evolutionary effects produced by such binaries if no compact objects are present and the cluster is composed entirely of normal stars, primarily those on the main sequence.

On this simplifying assumption, tidal-capture binaries will accumulate within the cluster core long before any binaries are formed by three-body processes, and will affect the core collapse when the number of binaries within the core is substantial—roughly a thousand, as we shall see below. The Fokker-Planck equation, while less reliable than for early evolutionary stages, when some 10^5 stars are interacting, should still provide a first approximation for these conditions. Hence, the effect of tidal-capture binaries can be evaluated with large-N cluster models, using either Monte Carlo techniques or numerical solutions of the Fokker-Planck equation.

We describe here the results obtained in one such model [5], which used the technique described in §4.1c for solving the Fokker-Planck equation numerically. As in other cluster models the three basic idealizations were: (a) an isolated cluster, (b) stars all of the same mass and radius, (c) a velocity distribution function dependent only on E and t, not on J. An additional basic approximation was the use of equation (6-42) for the rate of formation of binaries by tidal captures, with neglect of the complex phenomena associated with direct collisions. Similarly, direct collisions were ignored in considering subsequent interactions between these very hard binaries and either single stars or other binaries. Another simplifying assumption was that all binaries were formed with the same separation of $5R_s$ in their circular orbits. The cluster contained 3×10^5 stars initially, each of mass $0.7M_\odot$, distributed according to Plummer's model with R equal to 1.13 pc—see equation (1-17).

Cross-sections for interactions between binaries and single stars and between binary systems were taken from the results given in chapter 6. The appropriate distribution of released energy about its mean value was taken into account. Heating effects were included in the computations by removing the reaction products from the distribution function, $f(E,t)$, and inserting them elsewhere with the proper modified values of E.

The results obtained [5] are summarized in Fig. 7.3. The lower plot shows r_h and r_c (in units of R) plotted against t/t_{rh}, where t_{rh}, the reference relaxation time evaluated in equation (2-63), equals 2.3×10^8 years. The gradual development of core collapse is slightly accelerated by the presence of binaries; the time of collapse is about $14t_{rh}$, as compared with $16t_{rh}$ obtained from identical computations without binaries. In either case, the initial decrease of r_c results from build-up of the halo, as discussed in §4.2c. At a somewhat later stage, formation of binaries by tidal capture provides a loss of translational kinetic energy from the inner regions; §7.1 pointed out that the kinetic energy $\frac{1}{2}m_r V^2$, where m_r is the reduced mass

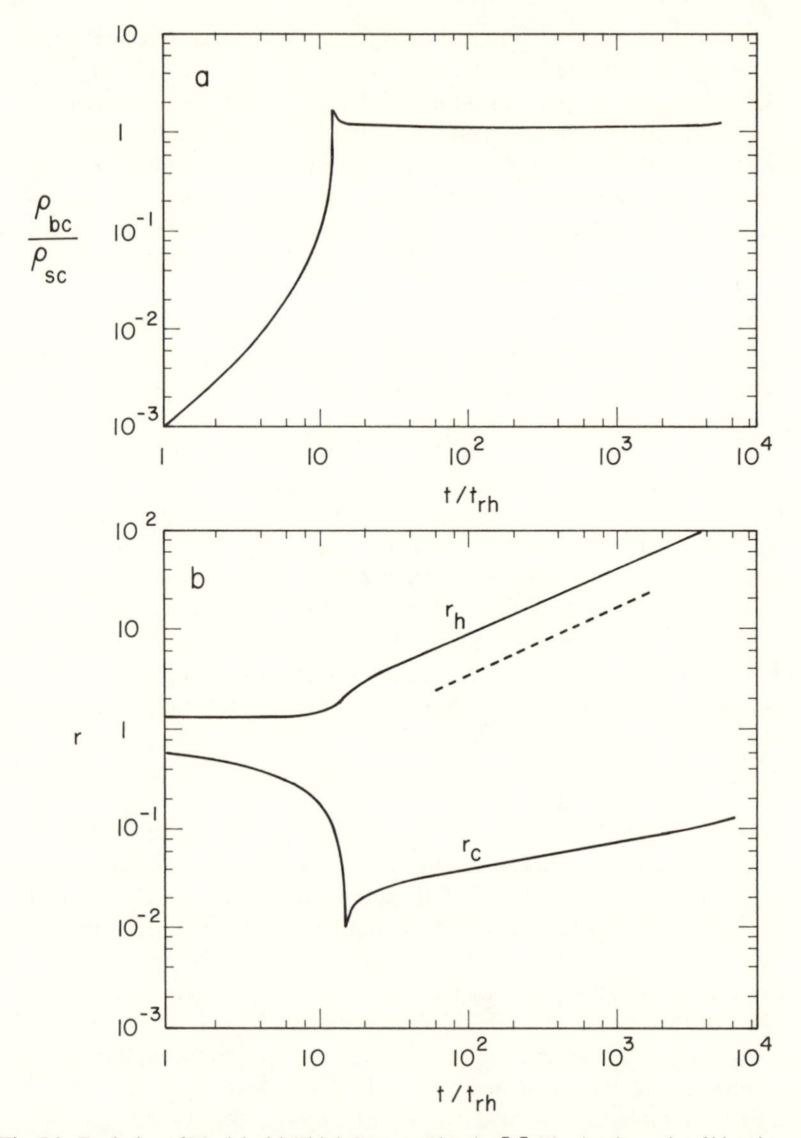

Fig. 7.3. Evolution of Model with Tidal-Capture Binaries [5]. The density ratio of binaries to single stars at the cluster center is shown in the upper plot a, while the half-mass radius r_h and core radius r_c are shown in the lower plot b. Both are indicated as functions of t/t_{rh}.

and V the initial relative velocity, is absorbed within the two interacting stars. This effect somewhat accelerates the collapse. The peak density reached in the collapse is about $10^8\ M_\odot\ \mathrm{pc}^{-3}$, at which time $v_m \approx 25$ km/s.

If binaries are present in the cluster initially, their tendency towards mass stratification greatly accelerates the collapse, as discussed in §4.2d. In the present instance this effect is small, since the newly formed binaries have the radial density distribution and the mean square random velocity appropriate for their increased mass. Since the formation rate is proportional to n_s^2, where n_s is the particle density of the single stars, n_b will initially be proportional to n_s^2, as required in equilibrium for particles of twice the stellar mass in an isothermal cluster. Equipartition of kinetic energies follows from the velocity distribution functions for $\mathbf{v}_1$ and $\mathbf{v}_2$, the two interacting stars, and consequently for $\mathbf{V}$ and $\mathbf{v}_g$, their relative velocity and the velocity of their common center of mass. To establish this result we write, from equation (2-10)

$$f^{(0)}(v_1)f^{(0)}(v_2) = \frac{j^6}{\pi^3} n_s^2 e^{-j^2v_1^2 - j^2v_2^2}$$

$$= \frac{j^6}{\pi^3} n_s^2 e^{-j^2V^2/2 - 2j^2v_g^2}. \tag{7-20}$$

Hence the distribution of v_g is Maxwellian with $\langle v_g^2 \rangle$ equal to half the mean square stellar velocity. Since the velocity distributions for V and v_g are separable, this conclusion is not affected by any variation of capture cross-section with V.

While the tidal-capture binaries are in equilibrium when formed, the remaining single stars are not, since binary formation produces a shortage of single stars in the central region. This effect is another aspect of the kinetic energy loss discussed above which, as we have seen, somewhat accelerates the contraction of the cluster core.

The upper plot in Fig. 7.3 shows the ratio of binaries to single stars, by mass, at the cluster center, again plotted against t/t_{rh}. As N_b increases, interactions will provide an energy source for the cluster. At about $t/t_{rh} \approx 10$ this energy source equals the rate of energy loss from binary formation, and the balance is available for heating the cluster. Gravitational heating is roughly twice the kinetic heating, and near the peak of the collapse, when there are 600 binaries and 710 single stars within r_c, the values of ρ_b and ρ_s are comparable in the core, and binary-binary interactions are important. When $\rho_b \geqslant \rho_s$ the mass stratification instability becomes significant, and this effect is presumably an important driving force for the collapse in its later stage. It is not clear whether the gravothermal instability contributes appreciably to the collapse rate; at the density peak, ρ_{cb}/ρ_{cs} is well below five, the value at which the gravothermal instability seems to dominate, according to Fig. 4.7.

While it is not easy to disentangle the various complex processes occurring in this model, the collapse evidently terminates, and the entire cluster expands; the subsequent evolution is discussed below. Similar model calculations have been made [5] with only gravitational heating included and also with only kinetic heating. The results are different in detail, with the peak density some ten times greater in both cases, but the collapse terminates only slightly later than before, at $t = 15t_{rh}$. These models indicate that core collapse will be terminated by tidal-capture binaries, if the cluster is composed of normal stars and if direct collisions can be ignored.

This result is not entirely conclusive in view of rather different results obtained from each of two separate Monte Carlo models. In one set of models [2], employing the dynamical method described in §4.1a, primordial hard binaries were assumed present, with up to half the mass in binaries initially. In a different orbit-averaged Monte Carlo computation [6], binaries were formed by tidal captures at the same rate used in the numerical solution of the Fokker-Planck equations leading to Fig. 7.3. In both of these Monte Carlo computations, the hard binaries stopped the collapse for a while, without much outward expansion, and the collapse then resumed. Both calculations made various approximations, especially in the cluster core. The dynamical Monte Carlo calculations did not extend very far into the collapse phase, and in the orbit-averaged model compact objects were concentrating in the core when the number of tidal-capture binaries reached its peak; perhaps it is for these reasons that their results differed from those shown in Fig. 7.3.

More important, the role of tidal-capture binaries in actual clusters is uncertain for two reasons. First the effect of direct collisions can certainly not be ignored. Half of the encounters which have been assumed to produce binary stars will involve collisions of stellar material, and will presumably lead to immediate loss of hot gas, leaving behind either a single star or two nuclei in a common envelope. Even if the two stars remain separate, their interaction with other stars and binaries will be strongly affected by direct collisions, which will much reduce the energy available for heating. Approximate numerical integrations for a single star interacting with a binary, whose identical stars are separated by $3R_s$, show [7] that tidal dissipation and direct collisions reduce the available heating energy by an order of magnitude, as well as leading to some stellar coalescence. These various effects may produce massive stars, which evolve rapidly and explode as supernovae. The mass lost by the cluster and the resultant gravitational heating may be much the same as for the tidal-capture binaries; the total mass involved is limited in either case by equation (6-42), ostensibly derived for tidal capture, but in fact including all the various objects formed by close stellar encounters.

A second major uncertainty in applying to clusters the theory of tidal-capture binaries concerns the fraction of the cluster mass in compact objects, such as degenerate stars, neutron stars and black holes. Some objects of this first group and all of the latter two are believed to be more massive than $0.7M_{\odot}$, the estimated maximum mass for luminous stars at present in globular clusters formed some 2×10^{10} years ago. As a result of mass stratification these relatively heavy objects will tend to concentrate in the core, where the fraction of normal stars will be substantially reduced. Some tidal captures of these objects by normal stars will almost certainly occur, as indicated by the presence of the X-ray sources discussed in §1.1. Encounters between such binaries and single compact objects will provide some heating, but the total heating by tidal-capture binaries will be much reduced if compact objects predominate in the core.

7.3 POST-COLLAPSE EVOLUTION

After the initial collapse of a cluster has been stopped, as a result of binary stars formed within the core, some expansion of the core seems likely. The newly formed binaries will tend to persist for a while and the heat source they provide would be expected to produce some expansion of the inner cluster region.

How far this expansion proceeds in actual clusters is uncertain. One simple possibility is that the system as a whole gradually approaches a quasi-steady state, with continuing expansion powered by the central energy source. Another more complex possibility is that this quasi-steady expansion is itself unstable, and that the dynamical state slowly fluctuates. A variant of this scenario is that the gravothermal instability soon reappears but produces core expansion rather than contraction, with the energy source at the center no longer active; the intermediate cluster regions contract to keep the total energy constant. Subsequent gravothermal collapse may then reoccur.

This final phase of evolution is important because many clusters have presumably already passed through the collapse phase with its brief very high transient density. Our understanding of the dynamical evolution will not be complete until the theoretical density profile $\rho(r)$ in the post-collapse phase is in agreement with observational data, particularly those obtained with high spatial resolution.

In view of our lack of definite information on post-collapse evolution of globular clusters, the present discussion will be brief, limited to homologous solutions and to a survey of some relevant model calculations.

a. Self-similar solutions

These solutions have the advantage that such quantities as M, r_h, t_{rh}, all show a simple dependence on time; the dependence of ρ and v_m^2 on the scaled radius can then be determined from a set of equations independent of time. We summarize the results for certain standard cases.

For a cluster with a tidal cut-off the Hénon self-similar solution discussed in §3.2b provides the earliest post-collapse model; the very high central density and its associated energy source were presumably formed in an earlier collapse phase. As we have seen in equations (3-21) to (3-23), the cluster disappears and $M = 0$ when $t = t_{rh}(0)/\xi_e \equiv t_1$. If we define $\tau \equiv t_1 - t$ as the time remaining to disappearance of the cluster, these equations yield

$$t_{rh} = \xi_e \tau, \qquad M \propto \tau, \qquad v_m^2 \propto \tau^{2/3}, \qquad r_h \propto \tau^{1/3}. \tag{7-21}$$

Fig. 3.2 (p. 00) shows the time-independent structure of this model, based on the assumption of an isotropic velocity distribution.

If the cluster is isolated and there is no tidal cut-off, ξ_e is much less. If mass loss is ignored and a self-similar solution again assumed, the cluster no longer disappears in a finite time, but expands indefinitely in the post-collapse phase. Again, this solution requires a source of energy, which may be assumed at the center. To obtain a self-similar solution we must now assume that $t_{rh} \propto t$, where t is the time since the start of self-similar expansion. With constant M, equation (2-63) and the Virial Theorem yield

$$r_h \propto t^{2/3}, \qquad v_m^2 \propto t^{-2/3}. \tag{7-22}$$

The structure of such a model has been determined [8] from a solution of the Fokker-Planck equation with an isotropic velocity distribution.

If mass loss is taken into account and M is assumed to vary as $t^{-\nu}$, the same procedure gives

$$r_h \propto t^{(2+\nu)/3}, \qquad v_m^2 \propto t^{-(2+4\nu)/3}. \tag{7-23}$$

If ν is determined from the mass-to-energy ratio for escaping reaction products, produced by the binary energy source, ζ in equation (3-4) is a large number [9]; from equation (3-6) we see that $\nu = 2/(3\zeta - 7)$ and is correspondingly small. The structure of such a post-collapse cluster has been computed with a perfect-gas model [9]. Mass is ejected from the cluster by the central energy source, providing gravitational heating throughout the cluster together with some gas inflow from outside the core; kinetic heating is all local at the center, contributing to an outward conductive heat flow. If the central energy source is attributed to binaries formed by three-body encounters, then for different assumptions $\nu \approx 0.013$ within a factor two.

In the central regions these post-collapse self-similar models neglect the finite size of the core and approach the solution for the singular isothermal sphere—see equations (1-28) to (1-30). A slight negative gradient of v_m^2 is superposed on this solution to drive the heat flow from the center.

If the central energy source results from binaries formed in three-body encounters, at the rate given in equation (6-37), a self-similar solution for the entire cluster in the post-collapse phase is possible [10] if the cluster is isolated and all stars have the same mass. Other assumptions required are that each binary, before ejection, releases an energy proportional to $\frac{1}{2}m[v_m(0)]^2$, the mean kinetic energy of single stars at the center, and that all this energy is deposited locally with no time delay; the resulting heating rate per unit volume, smoothed over time, then equals $\Gamma\rho^3/v_m^7$, where Γ is a known constant. The changes both of p and of cluster mass which result from ejection of reaction products are ignored. On these simplifying assumptions both r_c and r_h vary as $t^{2/3}$, in accordance with equation (7-22), and $r_c/r_h \propto N_c/N \propto N^{-2/3}$, where N is the total number of stars in the cluster.

The structure of such self-similar clusters and their rate of expansion have been analyzed with a fluid dynamical model; equation (3-38) has been used for the thermal conductivity, with $b \approx 0.45$ in accord† with §3.3b—see equation (3-48). For r between r_c and r_h, v_m is nearly constant and ρ varies about as $1/r^2$, the density variation found in equation (1-29) for a singular isothermal sphere. These detailed models are of particular use for the instability studies discussed at the end of this section.

b. Numerical solutions

We present next several detailed numerical models which include post-collapse evolution. The first of these is a Monte Carlo computation [6], using orbit-averaged changes in E and J. This approach, already referred to in §7.2b above, used a modification of the Hénon method, described in §4.1b. Most of the effects which might be important in actual clusters were included, such as formation of binaries both by tidal capture and by three-body encounters, stellar mass loss during early evolutionary phases, a galactic tidal cut-off and a wide distribution of initial stellar masses, from 0.1 to 3 $M_\odot$. The values obtained for the total bound mass M as a function of time are plotted in Fig. 7.4.

While this calculation is necessarily rather approximate, the general results are of interest. The total mass decreases nearly linearly, vanishing at

† The slightly different value of $f_1 \equiv 2b$ actually used [10] results from the different value taken for ζ.

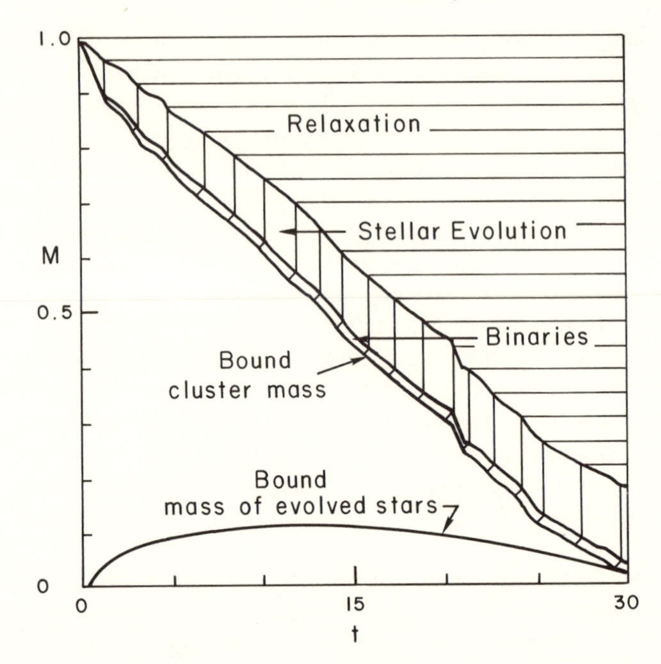

Fig. 7.4. Change of Mass with Time in Evolving Cluster. The total bound mass of a model cluster, computed with most known effects taken into account [6], is plotted against time in units of 10^9 years. Most of the mass is lost by velocity relaxation, giving stars enough energy to reach the tidal cut-off radius. The mass lost through stellar evolution and through interaction with binaries is shown, as is the total bound mass in stellar remnants.

3×10^{10} years, some 18 times the initial value of 1.7×10^9 years for t_{rh}, in remarkable agreement with the value of $22t_{rh}(0)$ obtained for the cluster life in the Hénon self-similar solution for a cluster with a tidal cut-off. The core collapse is stopped at $t = 12t_{rh}$ by binaries resulting from three-body encounters between the compact remnants of evolved stars, which dominate the core at this late stage. Some 1200 tidal-capture binaries are formed and disappear either by disruption or ejection, but most of their contribution to cluster heating occurs between 5 and 10×10^9 years (about 3 to $6t_{rh}$), long before the core of compact objects enters its collapse phase. As expected, the chief importance of stellar evolution during the cluster life is not so much the loss of mass from the stars and the cluster, but rather the formation of compact remnants which are each more massive than the luminous stars and which dominate the core during most of the cluster's life.

Another relevant model [5] is the numerical solution of the Fokker-Planck equation which includes binaries formed by tidal capture and which was described in §7.2b above. The plots in Fig. 7.3 show post-collapse evolution out to long times. Since the cluster was assumed isolated, escape of stars by relaxation was ignored, and the homology relations in equation (7-22) should be applicable; the constant v in equation (7-23) was found to be about 0.06, greater than for three-body binaries but still small. The dashed line in Fig. 7.3b shows the predicted variation of r_h as $t^{2/3}$, in good agreement with the model. Since r_c and r_h are different functions of t, when binaries in the core are formed by tidal capture, there is no self-similar solution for the expanding cluster as a whole.

The third model [11] is based on a solution of the fluid equations for a perfect gas. The cluster was begun as an $n = 5$ polytrope (Plummer's model), with 10^6 stars each of mass $1M_\odot$, and with $r_h = 5$ pc. An energy source was assumed equal to $C\rho^3/v_m$ per unit volume, where C was taken to be an adjustable constant. As noted in §7.1, the resultant energy is produced almost entirely within the core radius, r_c. Local deposition of the released energy was assumed; mass loss and gravitational heating were ignored. The resultantant values of ρ_c, the central density, are plotted in Fig. 7.5 as a function of t/t_{rh}. The three curves are computed for values of the constant C (expressed in dimensionless units) equal to 0, 10^4 and 10^6. For vanishing C the core collapse occurs as usual. For values of C between 10^3 and 10^5 the models show that the collapse terminates and that afterwards the core expands until ρ_c is only about four times its initial value, and then collapses again, with expansion and collapse alternating. Most of the time the cluster has an extended isothermal core. For higher C, the oscillations disappear, and the cluster slowly expands. These long-period oscillations have been found in several other investigations, based on Fokker-Planck models as well as fluid dynamic ones.

Late in the expansion phase of these oscillations the energy source is essentially turned off and the expansion is presumably powered by an instability of gravothermal type. The physical situation is that the energy source heats the inner region, which in consequence cools because of its negative specific heat—see §3.3a. Heat then flows from the warmer intermediate regions to the cooler inner ones, which continue to expand. When the expansion ends, the temperature difference between the intermediate and inner regions disappears, and the normal gravothermal instability produces a collapse of the core within the extended isothermal sphere.

Evidence for the appearance of instabilities in the post-expansion phase has also been obtained from a linearized stability analysis [10]. The configuration which was analyzed for the growth of infinitesimal perturbations was the self-similar expanding model described at the end of §7.3a. The results indicate that the expansion was stable for N less than about 7×10^3, overstable for N between 7×10^3 and 4×10^4, and unstable for greater N.

The increase of central concentration, measured by ρ_c/ρ_h, with increasing N may be physically related to the appearance of this instability. While these results do not apply directly to a post-collapse expanding cluster before it has reached a state of self-similar expansion, they verify the tendency of the post-collapse phase to instability, and can apparently account for large non-linear oscillations such as those evident in Fig. 7.5.

These models are all highly idealized, and the effects associated with more realistic assumptions remain to be explored. In addition, the manifold possibilities associated with non-linear oscillations need study. Quite apart from these unsolved dynamical questions, the post-collapse phase is uncertain in two other important respects. First, the complex effects produced by stellar collisions and, in particular, the evolution of the objects produced have not yet been explored; this is one area where dynamical evolution can depend on details of stellar evolution as yet unknown. Second, while compact objects may dominate the dynamics of the core, as pointed out at the end of §7.2a and b, their relative numbers are uncertain; while the fraction of cluster stars in degenerate dwarfs can be estimated, the fraction in neutron stars is very unclear, and as to the possible presence of black holes there is almost no information. Hence the possible dynamical

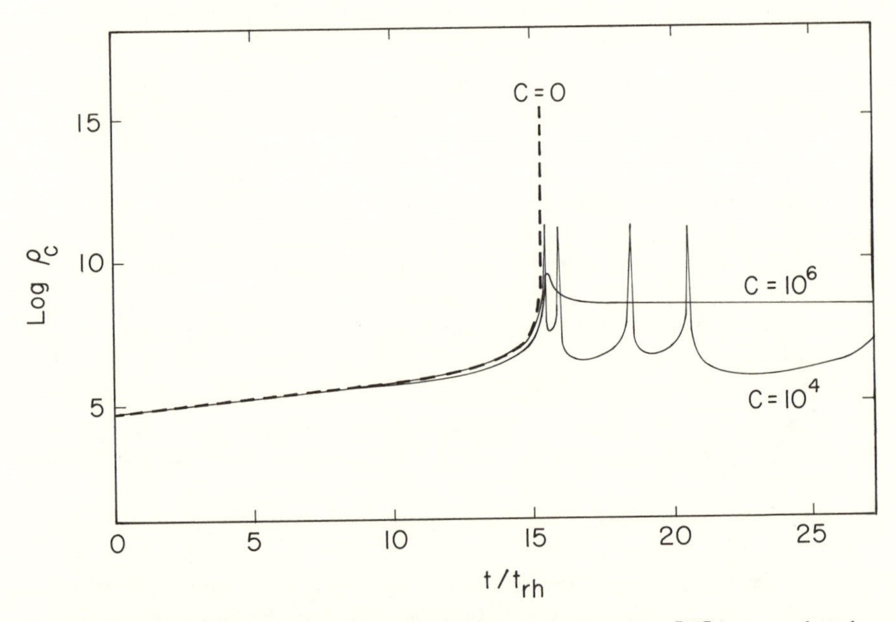

Fig. 7.5. Central Density in Evolving Cluster. The three curves show [11] $\log_{10} \rho_c$ plotted against t/t_{rh} for different values of the constant C assumed in the energy production rate for the central regions. The models are computed using the fluid equations.

importance of relatively massive compact objects in the cluster core, especially in tidal-capture binaries and in massive binaries formed by three-body encounters, either directly or through exchange processes, are quantitatively somewhat conjectural. A clear understanding of the final stages in a cluster's dynamical evolution is a goal for the future.

REFERENCES

1. S. Chandrasekhar and G. W. Wares, *Ap. J.*, **109**, 551, 1949.
2. L. Spitzer and R. Mathieu, *Ap. J.*, **241**, 618, 1980.
3. S.L.W. McMillan and A. P. Lightman, *Ap. J.*, **283**, 801 and 813, 1984.
4. M. Hénon, *Dynamics of Stellar Systems*, IAU Symp. No. 69, ed. A. Hayli (Reidel, Dordrecht), 1975, p. 133.
5. T. S. Statler, J. P. Ostriker and H. N. Cohn, *Ap. J.*, **316**, 626, 1987.
6. J. Stodółkiewicz, *Dynamics of Star Clusters*, IAU Symp. No. 113, ed. J. Goodman and P. Hut (Reidel, Dordrecht), 1985, p. 361.
7. S. L. W. McMillan, *Ap. J.*, **306**, 552, 1986.
8. M. Hénon, *Ann. d'Astroph.*, **28**, 62, 1965.
9. J. Goodman, *Ap. J.*, **280**, 298, 1984.
10. J. Goodman, *Ap. J.*, **313**, 576, 1987.
11. E. Bettwieser and D. Sugimoto, *M. N. Roy. Astr. Soc.*, **208**, 493, 1984.

List of Symbols

a		Acceleration, dv/dt; $a_i = dv_i/dt$.
		Semi-major axis of elliptical orbit.
A		Constant in equation (6-27) for differential cross-section, star interacting with binary.
b		Constant in equation (3-35) for conductive heat flow in a gas.
		Subscript referring to binaries n_b, N_b.
B		Constant in Maxwell-Boltzmann distribution law, equations (1-21), (1-22).
c		Subscript denoting value in cluster core, or value for a circular orbit, viz. J_c.
C		Subscript denoting quantities for clusters.
d		Subscript denoting galactic disc.
D		Density contrast $\rho(0)/\rho(R)$ in bounded sphere of radius R.
e		Eccentricity of two-body orbit, equation (2-1).
		Subscript denoting escape, viz. ξ_e, v_e, E_e.
E		Energy, usually per unit mass.
	E_T,E_c	Total energy ME and total energy of core.
	$E_n(v)$	Functions related to diffusion coefficients, equation (2-45).
f		Subscript denoting field stars.
	$f(\mathbf{r},\mathbf{v},t)$	Velocity distribution function; $f^{(0)}$, Maxwellian function, equation (2-10).
F	F_h	Conductive heat flow per unit area, equations (3-35) and (3-38).
	$F_n(v)$	Functions related to diffusion coefficients, equation (2-44).
g		Acceleration of gravity.
	g_m	Maximum $\|g\|$ in Z direction, well outside galactic disc.
G		Gravitational constant.
		Subscript denoting quantities for Galaxy, viz. M_G, R_G.
	$G(x)$	Function related to diffusion coefficients for Maxwellian distribution of field stars, equation (2-57).
j		Constant in Maxwellian distribution function, equation (2-11).
J		Angular momentum per unit mass.
	$J_c(E)$	Value of J for star in circular orbit with energy E.
m		Mass of a star or particle; m_j, mass of star of type j.
	m_r.	Reduced mass.
	m_t,m_f	Mass of test and field stars.

M		Mass of stellar system; M_C, M_G, mass of cluster and of Galaxy.
	$M(r)$	Mass interior to r.
	M_c	Mass of core within core radius r_c.
	M_j	Cluster mass in stars of type j.
n		Particle density; n_c, particle density $n(0)$ at cluster center. Polytropic index in a sphere, equation (1-16).
N		Number of stars, usually in a cluster.
	N_c	Number of stars within core radius r_c.
	N_b,N_s	Number of binary stars and of single stars in cluster.
p		Impact parameter.
	p_0	Value of p for 90° deflection in relative orbit, equation (2-5).
	$p(E,t)$	Phase space accessible to stars of energy E per unit dE, equation (2-78).
P	$P(E,J)$	Period of a cluster star with given E,J.
q	$q(E,t)$	Function related to $p(E,t)$, equation (2-84).
Q	$Q(E,J)$	Radial action integral, equation (2-92).
	$Q(x,y)$	Rate coefficient for increase of x by an amount y in star-binary interaction, equation (6-12).
r		Distance from center of cluster.
	r_t	Tidal cut-off radius beyond which stars without kinetic energy can escape from cluster.
	r_h	Radius containing half the cluster mass; r_{hP}, radius containing half the mass in projection.
	r_c	Core radius of cluster, at which projected surface density is half that at center.
	r_a,r_p	Radial distance at orbital apocenter and pericenter.
R		Distance; radius.
	R_G	Distance from center of cluster to center of Galaxy.
	R_p	Value of R_G at perigalacticon.
	R_s	Radius of a star.
s		Subscript referring to single stars.
S		Total entropy of cluster.
t		Time. Subscript denoting test stars.
	t_r	Relaxation time, equations (2-61), (2-62); t_R, reference relaxation time, equation (2-75).
	t_{rh}	Half-mass relaxation time in cluster, equation (2-63).
	t_{rc}	Value of t_r at cluster center.
	t_d	Dynamical time r/v_m; t_{dc}, t_{dh}, value of t_d for $r = r_c$ and $r = r_h$.

	t_{sh}	Shock heating time at $r = r_h$, equation (5-30).
T		Kinetic energy of stellar system.
u		Fluid velocity; mean velocity of stars in unit volume.
v		Velocity of a star.
	v_m	Root-mean-square random three-dimensional velocity.
	v_r, v_t	Velocity components parallel (radial) and perpendicular (transverse) to **r**.
	v_e	Velocity of escape; v_{es}, escape velocity from star.
V		Relative velocity, usually at infinite separation.
	V_c	Critical value of V giving zero total energy for star interacting with binary, equation (6-11).
	V_p	Relative velocity of cluster relative to Galaxy at pericenter, $R_G = R_p$.
w		Kinetic energy of relative motion at infinite separation of two bodies.
W		Total gravitational energy of a system.
x		Net binding energy of a binary star, equation (6-2).
		Ratio of velocity to $(2/3)^{1/2} v_{mj}$; equals jv.
	x_e	Distance from cluster center to Roche surface, measured toward galactic center.
X	X_0	Dimensionless central potential in cluster, equation (4-15).
y		Change of net binding energy of a binary in interaction with passing star.
z		Distance of star from cluster center, in direction perpendicular to galactic plane.
Z		Distance above galactic plane; Z_C, value of Z for center of cluster.
β		Measure of ratio, duration of perturbation to orbital period, equation (5-27).
		Constant in Maxwell-Boltzmann equation, $\beta_j = B_j/m_j$, equation (6-8).
	β_r	Value of B/m for star-binary encounters, equation (6-22).
Γ		Constant factor in diffusion coefficients, equation (2-13).
Δ		Relative change of binary binding energy in a stellar encounter, y/x.
		Finite increment.
	$\Delta v_{\parallel}$	Component of $\Delta \mathbf{v}$ parallel to initial $\mathbf{v}$, equation (2-19).
	$\Delta v_{\perp}$	Component of $\Delta \mathbf{v}$ perpendicular to initial $\mathbf{v}$, equation (2-18).
	$(\Delta E)_{\mathrm{Av}}$	Shock heating rate averaged over stars with a particular z or r, equations (5-14) and (5-36).
ε	$\varepsilon_1, \varepsilon_2$	Mean change and rms change of E in one orbit.

ζ		$d \ln E_T/d \ln M$, equation (3-4), and $d \ln E_c/d \ln M_c$, equation (3-30).
θ		Angle.
		Dimensionless potential in isothermal sphere, equation (1-25).
κ		Scale distance in isothermal sphere, equation (1-24).
λ		Expansion parameter, equations (5-16), (7-17).
Λ		Value of $p_{\max}/p_0$, equations (2-14) and (7-3).
ξ		Dimensionless radius in isothermal sphere, equation (1-24).
	ξ_e, ξ_c	$t_r d \ln M/dt$ and $t_{rc} d \ln \rho_c/dt$, equations (3-3) and (3-29).
ρ		Density, usually smoothed density in cluster.
	ρ_c	Density $\rho(0)$ at cluster center.
	ρ_G	Mean density of Galaxy within sphere of radius R_G.
σ		Surface density of cluster.
		Cross-section for interaction, equation (6-15).
	σ_c	Cross-section for capture of star by binary.
	σ_{dss}	Cross-section for dissociation of binary by star, equation (6-26).
	σ_{dsr}	Cross-section for disruptive collision between two binaries, equation (6-33).
τ		Time remaining until collapse, $t_{\mathrm{coll}} - t$.
ϕ		Gravitational potential.
	$\phi_d(Z)$	Potential produced by galactic disc.
Φ	$\Phi(x)$	Error function, equation (2-56).
χ		Critical quantity for mass stratification instability, equation (3-54).
		Angle of deflection in relative two-body orbit, Fig. 2.1.
ψ		Dimensionless tidally distorted cluster potential, Fig. 5.1.
ω		Angular velocity.
$\langle X \rangle$		Average value of X.
	$\langle (\Delta v_i)^n \rangle$	Average of $(\Delta v_i)^n$ over all encounters experienced by a test star in a unit time; i.e., diffusion coefficient.
	$\langle \Delta E \rangle_{\mathrm{orb}}$	Average of $\langle \Delta E \rangle$ over an orbit for a given E,J, equation (2-91).
	$\langle \Delta E \rangle_V$	Average of $\langle \Delta E \rangle$ over volume of accessible phase space for a given E, equation (2-72).
	$\langle \sigma \Delta \rangle$	Integral of $(d\sigma/d\Delta)\Delta\, d\Delta$ over Δ, equal to $1/(n_s V x)$ times rate of energy transfer from binary to single stars, equation (6-28).
*		Subscript denoting, in a self-similar solution for an evolving cluster, a function of scaled radius, equations (3-28).

Index